SURVIVING ENERGY PRICES

SURVIVING ENERGY PRICES

Peter C. Beutel

DISCLAIMER

While this book discusses the use of futures, options, caps, and swaps, it is in no way to be considered an offer to buy or sell these or other instruments. These instruments involve substantial financial risk and should be used with the greatest caution and with the supervision of a trained professional. Neither the author nor the publisher is willing to assume any liability for losses that may be incurred through the use of any financial instruments. All the information in this book is believed to be reliable, but neither the author nor publisher can guarantee its accuracy. The availability of new and potentially better instruments may change our view on the efficacy of strategies outlined in the most recent copy of this book.

Copyright ©2005 PennWell Corporation
1421 South Sheridan Road
Tulsa, Oklahoma 74112
1-800-752-9764
sales@pennwell.com
www.pennwell.com
www.pennwell-store.com

Managing Editor: Marla Patterson
Production Manager: Julie Shank
Book Designer: Wes Rowell
Cover Designer: Ken Wood

Library of Congress Cataloging-in-Publication Data
Beutel, Peter C.
  Surviving energy prices / Peter C. Beutel.
    p. cm.
  Includes index.
  ISBN 1-59370-042-3
  1. Petroleum products--Prices. 2. Hedging (Finance) I. Title.
  HD9560.4.B46 2005
  332.64'4228--dc22

                                                                2004024888

Printed in the United States of America
1 2 3 4 5   09 08 07 06 05

DEDICATION

My special thanks to Grant M. Holroyd for his painstaking work in editing and formatting this work.

All my love and thanks to my mother and father.

CONTENTS

SECTION I–THE PROBLEMS

SECTION II–THE SOLUTIONS:
DIFFERENT METHODS OF HEDGING

PREFACE

Energy prices move constantly and often dramatically. As I am finishing this book, gasoline and crude oil prices have reached their highest prices ever. The economy is growing today. But, ultimately, an improving economy could push the prices for energy so high that they cripple or maim the growth that caused them in the first place. Since the 1970s, energy prices have had a very direct impact on the American economy. That is expected to be the case, at least to some degree, in future years.

I have written this book in the hope that it can help educate a new generation of business people who have discovered, to their great horror, that changing energy costs can have a gargantuan effect upon their company's bottom-line. Even if one never intends to hedge, one should be aware of the elements that can affect the bottom-line.

My company, Cameron Hanover, publishes a daily report on changing energy market conditions. Those who would like to learn more may call me directly at (203) 801-0771, or visit our Web site at *www.cameronhanover.com.*

INTRODUCTION

As the new millennium begins, it may be appropriate to reflect a little on the history of oil and oil trading. Approximately 150 years ago, oil was an *alternative* source of energy, like wind power and solar energy are today. At that time, coal was primarily used for heat, and whale oil was used for light. Nobody had much use for petroleum, and there was no economic incentive to market it. Everyone knew it was there. It had been named *petroleum,* or "oil of the rock," by the ancient Greeks. Until the price of whale oil became high enough to encourage the search for another fuel, it just was not economical to bring the oil out of the ground.

The change in market pressures came rather suddenly. Shortly before the Civil War, whales became scarce, because they had been unwittingly hunted to the brink of extinction. Whalers first thought that they were just having a bad year, but as one bad year followed another, it became apparent that whales were not the renewable resource that people had once thought. As whale oil became increasingly scarce, prices increased dramatically. When prices reached a level high enough to make petroleum economically viable, people became willing to replace whale oil. This launched a new industry and created a new class of entrepreneurs.

First among these was John D. Rockefeller—the Bill Gates of his era. He saw that oil could replace other fuels more efficiently and less expensively and saw a growing industrial nation thirsting for energy. His company, Standard Oil, established

Kerosene is one of several distillates that is refined from crude oil. It is similar to jet fuel and is used for lighting and space heating needs. It is also used to cut heating oil or diesel to give it a lower pour point.

itself in this new market by promising a consistent quality of kerosene for lighting purposes: a *standard* grade.

The demand for oil expanded with each new technological innovation. The invention of the automobile created a burgeoning demand for a new oil-based fuel. Henry Ford applied Standard Oil's underlying principle of sameness to cars. He made his cars standard by building them on an assembly line. That made them less expensive to build and, in turn, to buy. This created an entirely new level of popular interest. Demand increased as prices fell. That was true for both automobiles and for oil.

Neither of these men could have predicted the American love affair with the automobile. Henry Ford's dream of every working person being able to buy one of his cars was fully realized during his lifetime. It set a standard of car ownership that became deeply imbued in American society. As everyone aspired to own an automobile, the demand for oil grew exponentially, and it is still growing today.

The growth in demand for oil did not stop with the car. Other innovations also required a mode of power that increasingly required the further refinement of crude oil into specific types of fuel.

John D. Rockefeller's Standard Oil Company became the first arbiter of oil prices. There had been an active futures market in 1890, but Mr. Rockefeller eliminated all trading by offering to buy and sell an unlimited amount of crude oil at 50¢/barrel (bbl). By taking both sides of the market at the same price, he killed any likelihood of prices moving, and that drove speculators away, melting market liquidity. In effect, he became the market all by himself. It was not until the late 1970s that an active trade in oil futures became a part of the oil trading landscape again.

Just as Bill Gates learned with his Microsoft Corporation, the U.S. government is wary of one company having too much control over its market. In 1911, the government broke up Standard Oil into competing companies. The Texas Railroad Commission set prices in the intervening time before a coalition of companies, known as the Seven Sisters, became the new arbiters of oil prices. It consisted of Standard Oil's successors and a couple of other large international oil companies. They kept prices low, demand high, and together effectively ran the entire oil business. Prices were not just stable. They were static.

After World War II, as many oil-producing countries became independent, resentment grew over corporate control of oil prices. These countries formed a cartel known as OPEC, the Organization of Petroleum Exporting Countries. The group came into its own during the 1960s, and by the 1970s, it was a powerful political force. After the Arab-Israeli War in 1973, the Arab oil-exporting nations banded together to punish Israel's ally, the United States, by imposing an oil embargo. This was the first oil price shock the United States would experience.

A second price shock followed in 1979, when Iranian religious nationalists overthrew the monarchy of the Shah while he was seeking medical treatment in the United States. The Iranians wanted him extradited to Iran to stand trial. They took Americans hostage in an attempt to trade for him, and they boycotted oil sales to the United States.

For the second time in a decade, Americans waited in long gasoline lines and experienced a political energy crisis. This crisis shackled the U.S. economy and led to a combination of high inflation and sluggish economic growth. It impacted the fuel markets in the same way that the whale oil shortage 120 years earlier affected the cost of fuel and paved the road for oil use. This newer price shock led to fuel conservation and a search for alternative fuels.

OPEC paid dearly in the 1980s for its experimentation with higher oil prices, as oil prices collapsed from new production and lower consumption. It took years for demand and prices to recover from the shocks of the 1970s.

One of the major changes that accompanied the oil shocks of the 1970s was the return of active trading in oil markets. In 1978, the New York Mercantile Exchange (NYMEX) introduced trading in heating oil futures, and in the 1980s, it added crude oil and gasoline futures. Futures did not create volatility, but they certainly did not end it.

For a host of other end-users, the volatility that followed the twin oil shocks of the 1970s became a major source of agony. Municipalities, heating oil distributors, diesel resellers, railroads, airlines, trucking companies, factories, gasoline and diesel marketers, refiners, and producers all felt the impact. It was as if there was suddenly a big green monster on the front porch, and it was sucking money out the front door. On some days it brought flowers, and the money went whooshing back in. However, there was no way of telling what it would do on any given day, or for that matter, during any given day.

Of course, there never really was any monster, but that's the way many distributors and end-users felt as they tried to deal with the volatility. This new unpredictability in oil prices might just as well have run a vacuum hose straight into these companies' bank accounts.

Most companies had no idea how to deal with the rapid changes they saw every day, and their customers were becoming confused and angry. Consumers saw prices either rising or falling on their bills and at the pump. They no longer felt confident in the price stability of their primary sources of fuel.

Consumers switched to natural gas or bought their oil from COD (cash on delivery) suppliers, equipped with little more than a truck and a phone. There was no way to avoid higher gasoline prices. Cold winters did not just bring good sledding conditions any more. They could ruin a company or a family's budget. Companies providing or using fuel tried to absorb higher costs when prices were higher, and they watched in shock as their profit margins got increasingly smaller.

Consumers and their suppliers were not alone in the dark. Diesel users, like railroads and trucking companies, and jet fuel users, like airlines, had huge fuel needs whose costs were impossible to predict reliably in budget forecasts. Unexpectedly large fuel costs could mean the difference between running a company profitably or going into the red. One year, oil price volatility might mean a brace of yachts for the directors and Christmas bonuses for all. Another year, it could bring layoffs and cost-cutting measures for the company just to stay in business. Clearly, energy input costs needed to be controlled.

In the last 20 years, rampant price fluctuations have forced everyone buying, selling, or using oil to reconsider the way they do business. In 1990 Saddam Hussein invaded Kuwait, and 5 million barrels per day (bpd) of oil were removed from the market. Oil prices tripled in five months and eventually led to a recession in the West. The higher oil prices of 1990–91 led to lower demand and also encouraged fresh exploration and drilling. This helped to lower oil prices and laid the foundation for a period of exceptional Western economic growth in the middle and late 1990s.

The 21st century has ushered in the new economy of the Information Age, transitioning from the previous Industrial Age. Nevertheless, it would be a mistake to downplay the continuing role of oil. As long as there has been a human race, it has struggled for two basic necessities: food and fuel.

Those two remain central to human existence. The technological advances of civilization have certainly changed the procurement process, but these basic needs remain constant. There is an energy input involved in every process, and one can expect that a new generation of companies will discover just how important energy costs are to their businesses.

At the end of the 1990s, there was a large school of thought that energy inputs were no longer central to the economy. America had come off the largest stock market boom in its history. And, as is generally the case with these events, no one saw it ending.

However, there was an insidious development in the economy—oil prices broke above $30/bbl and stayed there. The U.S. economy has never grown during periods of high energy costs. That is one of the rediscovered truths of the early 21st century. High energy costs impede, retard, and cancel benefits elsewhere in the economy. Every time oil prices have exceeded $30/bbl in the last 30 years, there has been a recession, inflation, or both in the 12–18 months following the price advance. As this book goes to press, this cycle could occur again, in 2004 or following.

This book is designed to teach the basics of hedging crude oil, natural gas, and refined petroleum products. It specifically addresses the new realities of protecting against adverse price moves in crude oil, heating oil, diesel, gasoline, and natural gas, and their effects upon the businesses using them. It discusses *futures, options, wet barrel programs,* and touches on *swaps,* and it will discuss the methods that buyers and sellers are using to stay profitable. It demystifies *margin calls, technical analysis* of commodity price patterns, and the unique set of *fundamental* factors that can affect price movements. It discusses ways to use storage to lower costs, and it talks about *fixed price* and *capped price* programs, which can be offered to customers to keep them from moving to a competitor. It also warns about the problems of *basis risk* between various locations.

The world keeps changing, and there is nothing new in that. This book was written in hopes that it will help readers to keep up with the changes that have occurred in the business of buying and selling energy. The seasonal tables can give readers a leg up in terms of planning a yearly strategy or trading and hedging campaign. The explanations have been kept simple for those just starting to familiarize themselves with these issues. Even so, the seasoned trader will find a few golden nuggets of information that will prove useful. This book includes all of the important facets of trading and hedging, with hopes that it clarifies what is really going on in this business today. So it is time to get down to brass tacks, or more appropriately, cash racks.

SECTION I

THE PROBLEMS

For most of the 20th century, there was remarkable stability in oil prices. No one needed to hedge, because the risks were minimal. Then came the 1970s, a period of political and financial uncertainty that effectively changed the entire complexion of the oil business. Traders were forced to discover ways to deal with the new realities through trial and error.

With the beginning of the new millennium, the changes in energy prices have become even more dramatic. Crude oil prices were just above $10/bbl in December 1998. By May 2004, crude oil prices were more than $42/bbl. Companies using or supplying energy have encountered a new range of problems that have been the direct result of these changes in energy costs.

Fortunately, there are ways of dealing with changing energy prices. The following chapters present the methods that have been distilled during more than a quarter century of experimentation.

The New Volatility

Up until 1973–74, oil prices were extremely stable. If the price of heating oil, diesel, or gasoline changed more than 1¢–2¢/gallon (gal) over the course of a year, it was considered extreme. However, with the Arab Oil Embargo of 1974, all that changed. Prices started moving, and they never looked back.

With the next oil shock in 1979–80, oil prices quadrupled in a very short time. The conservation movement *There are three major bench-mark crude oils in use in the world today. In the Western Hemisphere, West Texas Intermediate, a light, sweet crude, is the benchmark against which other crudes are priced either at a discount or a premium.*

In the Far East, Dubai is the benchmark. In Europe, Brent Blend is used as the benchmark.

was born, and the world economy went into a recession, largely brought on by the inflationary effects of higher oil prices. The next stage was in the opposite direction. The high prices of 1979 and early 1980 sowed the seeds of their own reversal, and the recession and conservation they spawned exacerbated low prices. By April 1, 1986, despite OPEC's best efforts to stop the hemorrhaging, prices had been cut by a factor of four. Crude oil dropped from $39.80/bbl to $9.75/bbl basis West Texas Intermediate, the Western Hemisphere benchmark.

This was the first full cycle of oil price volatility, and heating oil prices dropped from $1.05/gal in 1980 to less than $0.3000/gal in early 1986. Other price shocks followed. In the summer of 1989, heating oil prices fell to a low of $0.4440/gal. By December 1989, which was the coldest December of the last century, prices were around $1.10/gal. In June 1990, heating oil prices reached a low of $0.4720/gal, and crude oil prices touched a low of $15.06/bbl.

Shortly thereafter, on August 2, 1990, Saddam Hussein invaded Kuwait, and heating oil prices reached $1.0850/gal by October. Crude oil prices on the NYMEX reached $41.15/bbl. Volatility had reached new levels, and prices had demonstrated convincingly that they were capable of doubling or halving in six months or less.

The same boom-and-bust cycle has occurred again recently in the United States. Crude oil prices dropped below $11/bbl in December 1998. By late February 2003, they stopped 1¢ short of touching $40/bbl. Heating oil prices established a high of $1.31/gal in the winter of 2003, on a combination of cold weather and war fears (leading up to the invasion of Iraq).

A combination of events pushed prices to the highest levels ever in 2004. Crude oil prices were propelled by a general strike in Venezuela, sabotage and terror in Saudi Arabia and Iraq, recurring problems in Nigeria, and surging demand in the United States, India, and China. American businesses and consumers watched their money evaporate at the pump and braced for a winter of discontent over surging heating costs. Billions of dollars that had been earmarked for a general economic recovery were spent on gasoline, diesel, jet fuel, and heating oil. Businesses and consumers scrambled to survive the energy price increases. Although the economic costs have yet to be reckoned, it is obvious that nothing has changed in terms of price volatility.

A step back in time helps put this in proper perspective. After having paid reasonable prices in the mid-1980s, many consumers were unable to understand the higher prices except in the context of a large conspiracy. This conspiracy, consumers often wrongfully believed, included gasoline retailers, heating oil distributors, railways, airlines, trucking companies, and other resellers and end-users. Each was forced to pass along higher energy input costs.

Consumers reasoned that the weather turning bitter or Saddam Hussein's invasion of some faraway place was not an excuse for ridiculously higher prices. A heating oil distributor should be able to supply the fuel to heat a home without charging an arm and a leg. And refiners should be able to get crude out of the ground and to pumps as gasoline without breaking consumers' backs. Some consumer groups feared that the big oil companies were working hand-in-glove with retailers and distributors to milk consumers dry.

This was not the case, but as with the price gyrations of the underlying fuel, it was the perception rather than the reality that made the biggest difference to consumers. Back-to-back winters with extreme price increases in 1989–90 and 1990–91 represented the breaking point for many. Heating oil distributors who needed to increase prices in order to reflect their own higher costs, lost customers, or profit margins. The perception that everyone involved with the supply of oil was striking it rich, at the expense of abused consumers, led to a general erosion of the goodwill built up over many years by suppliers.

This was clearly not good for gasoline retailers, heating oil distributors, or for transportation and freight companies that used fuel. They were caught between their suppliers and the consumer, while the market tightened its grip. Prices advanced because no one had anticipated a December as cold as the one in 1989, and because traders feared that Saddam Hussein could invade Saudi Arabia and dictate the price the world would pay for oil. Prices came back down fairly quickly in both cases, but the damage was done. Heating oil consumers had started switching to natural gas, or they shopped around for each week's lowest price. They might not know the delivery driver's name any more, but it no longer seemed worth the price. This same creeping cynicism was extended to the price increases that forced other companies to pass along their higher energy costs. Consumers lost confidence in every aspect of the oil supply chain.

The difference between home heating oil and diesel is the amount of sulfur and cetane in each product. Cetane is to distillate (which includes both heating oil and diesel) what octane is to gasoline, more or less. Heating oil has 0.2% sulfur, hence the term number two oil, while diesel has 0.05%. Heating oil has a 40 cetane, while diesel has a 45 cetane. The two items are close enough in their specifications to make it a matter of blending up or down to turn one into the other.

Through most of the 1990s, prices were relatively low and stable. By 1998, the effects of years of conservation efforts and fuel switching had changed the dynamics of supply and demand dramatically. OPEC was forced to cut production repeatedly to keep prices from plummeting. It did not work. Light, sweet crude oil on the NYMEX had fallen to $10.35/bbl by December 1998, which was lower in real terms, even after years of relatively mild inflation.

The low prices decimated U.S. and Canadian exploration and drilling expenditures, and came at a time when natural gas and electricity were relatively cheap as well. Funding dried up for any further expansion of energy-producing fuels or technologies. At the same time, sports utility vehicles became a larger proportion of the vehicle fleet. With U.S. consumers enjoying the lowest real costs for gasoline in years, they reignited a long dormant love affair with large and fuel-inefficient vehicles. The economy grew in 1999 and into 2000 on the back of low energy costs.

As had occurred before, the low real costs of energy led to profligacy. Demand grew steadily, but fresh supplies were not coming to the market. There is a lead time involved in finding and producing crude oil or natural gas. Because development expenditures had been slashed along with the low prices in 1998 and 1999, new supplies were not keeping pace with surging demand.

Prices per barrel ended the 20th century in the high $20s and low $30s. They had tripled in just about a year, and they were to get worse as the year progressed. Oil and gas companies increased their exploration efforts, but the waiting time hurt them. By autumn 2000, per-barrel prices were in the high $30s and were showing signs of moving higher. With less than seven weeks until a national election, President Clinton decided on September 22, 2000 to release oil from the Strategic Petroleum Reserve (SPR).

The additional oil from the SPR broke the speculative bubble, but it did nothing to produce the extra heating oil that the market needed badly before the winter. It was no longer a problem of enough crude oil but a problem of not enough refining capacity. Crude oil prices fell $6–$7/bbl but consolidated at more than $30/bbl.

Since then, the volatility has only gotten worse. In February 2002, crude oil prices fell below $17/bbl. But by autumn, they were back up to more than $30/bbl again. In early 2003, a strike in Venezuela and fear of another war with Iraq pushed oil prices to their highest levels in 12 years. And the economy felt the hit from higher energy costs across the board. The stock markets had crashed, consumer confidence had eroded, and the year came on the heels of the worst December retailing season in 30 years.

These economic effects were in large part the results of higher energy costs. In fact, American consumers were paying roughly $300 million more each and every day for their petroleum needs in January 2003 than they had been in February 2002. By late winter of 2003, low inventories and OPEC production cuts had pushed prices to more than $36/bbl, even as the economy was starting to show signs of recovering.

When prices are high, end-users who need to work within pre-established budgets suffer. Budgets have already calculated fuel use and have provided a certain amount for its purchase. When prices have sky-rocketed, those budgets have flown out the window, and hapless fuel managers have been blamed for events seemingly beyond their control. This process hurts companies, consumers, and local governments trying to plan ahead for fuel use.

Companies and local governments have a genuine problem when it comes to deciding whether to pass on higher costs to customers and tax-payers or to absorb those costs themselves. In some cases, it is not worth losing customers or antagonizing voters because of price volatility. Many entities have absorbed the difference in order to maintain goodwill, but it has become obvious after several episodes that this is hardly a good way to run a business or a town. The profitability of companies and viability of municipalities cannot be left to the vicissitudes of the markets.

Fortunately, there are better ways to handle price uncertainty, and the following chapters are devoted to explaining them in detail. In today's world of extraordinary price volatility, it is dangerous to leave fuel prices unhedged.

Understanding Hedging

In any discussion of *hedging* with people in the energy business, there are sure to be those who believe it is a complete waste of time. "I lost $100,000 I could have taken to Las Vegas," says one. "My broker got me speculating in pork bellies," complains another. Still another will talk about the time she hedged and missed the opportunity to make an extra nickel on her margins. Some executives compare hedging unfavorably with gambling, conjuring up images in their minds of fast-talking New Yorkers wearing white shoes, red suspenders, and pinstriped suits.

The fast-talkers and gamblers are out there in the market. However, for those companies that are directly affected by any changes in the market, inherently by their physical positions, hedging can save them. They do not want to become the subject of future stories, like those of the unhedged distributors who sold oil at cost in the winters of 1989–90, 1990–91, or in 2002–03. Neither do they want to be like the wholesalers holding 2 million bbl of heating oil, who suddenly had to sell them at a 10¢/gal loss.

Differentials can explode because of extreme temperatures or because of refinery problems in one region. They almost always reflect some combination of time, or an urgent need for material right away, and location, other than the main delivery points in New York Harbor.

Hedging is not the same as speculating, when it is done properly. And in today's market environment, holding an unhedged position in any energy commodity *is* speculating. There is no difference between the risks taken by a conservative businessman holding unhedged inventory and a wild-eyed speculator. Their risks and exposures are precisely the same. Some readers may wonder how avoiding futures or options could in itself be speculating. This is quite easily the case, as it turns out.

Futures and cash prices tend to move in tandem. Cash diesel, gasoline, or jet fuel barrels sitting in storage are going to rise or fall in value along with futures. The same is true for crude oil, any other refined product, or natural gas. During the 1980s, cash prices and futures converged so completely that they are now largely indistinguishable. Most cash prices are given as a differential to futures today. If futures prices advance a dime, it is almost certain that cash prices in New York Harbor, at Cushing, Oklahoma, or at the Henry Hub in Louisiana will also advance very close to 10¢. They may advance by 9¢ or 11¢, but cash prices will certainly move in the same direction by a comparable amount.

A distributor waiting to buy heating oil, or a railroad waiting to buy diesel, might have to pay 3¢/gal more when futures rise 10¢/gal. There are often steep *prompt premiums*, but they rarely occur without there first being a move in the underlying futures price. (Prompt premiums are explained in more detail in a later chapter.) In general, there are changing degrees of urgency to get product that can be picked up right away, or on a prompt schedule. But there is almost always a certain amount of directional movement that both markets will share in common.

By waiting to buy, would-be buyers are in effect *short* product. They are speculating that prices will drop, and that they will be able to buy fuel that is less expensive. And in that sense, they are gambling, either with their customers' goodwill or with their company's profit margin. If prices advance, it is going to come from one of those two pockets—either the company's or the consumer's.

Those who waited until October 1990 to buy that winter's heating oil supply found this out the hard way. They could have paid less than $0.50/gal in June but instead paid more than $1/gal by October. Someone had to pay the difference, whether it was the customers paying $1.50/gal and swearing that next year they would use natural gas or the distributor selling its oil at cost. Either way, the business suffered.

Municipalities that did not hedge their heating needs for the winter of 2002–03 were in the same position. Over the summer, natural gas prices were under $3/million British thermal units (mmBtu). During the heart of that winter, natural gas prices were more than $5/mmBtu. By the end of February, they had traded up to almost $12/mmBtu. On top of that, many municipalities found that tax receipts were lower as they started 2003. Cities and towns nationwide were struggling with higher energy costs and fewer funds with which to pay them.

Effective hedging is nothing more or less than locking in an operating profit. By being willing to forego the windfall profits that might come with lower prices, a hedger is guaranteeing that he will not have to lower margins or raise prices in an advancing market. The object here is not to make speculative profits but to protect operating profits, year in and year out. For businesses that went unhedged, it is certainly possible that the winters of 1991–92 and 1992–93 could have made up for the winters of 1989–90 and 1990–91. It also is possible that the next few winters will make up for the most recent one. But this is hardly a sensible way to run a business. Although heating oil and natural gas are given here as examples, the theories and principles apply equally to all energy products.

Banks do not feel that leaving inventory positions unhedged is a good way to run a business, either. In fact, hedging is one of the words a heating oil distributor or fuel oil purchasing manager can say that will make a banker smile. When bankers lend money to companies, their greatest concern is that the debt can be serviced through operating profits. Bankers do not want to get sudden calls to cover cash shortfalls just because the price of an underlying product has gone up. They like to know what the costs associated with fuel procurement are going to be in advance. Bankers hate surprises, and hedging can eliminate surprises.

Banks are not in the business of financing speculators, which is why the government has to give tax breaks to investors in oil exploration or *wildcatting*. Bankers do not want the possibility of owning a dry hole in Oklahoma, and they feel the same way about any other business. They want to lend money to steady businesses that can reliably generate profits. It is important to remember that what a business calls *profits*, a banker considers *the ability to repay*.

Hedging can give a business the edge of financial stability that can spell the difference between a yes or a no in a discussion with a banker. This is true whether the hedging is accomplished through futures or options, wet barrel deals with suppliers, or swap arrangements with financial institutions. It is a sign of financial sophistication that shows a banker that the borrower understands the business risks. Since the banker is such an important partner in any business, maintaining a good relationship is a primary concern. The understanding of this becomes more critical in those situations in which the company is small or dependent upon the bank for its operating capital.

Many companies rely upon their suppliers as their *de facto* bankers. Rather than paying within 10 days and receiving a 1%–2% discount, many companies pay on a 30-day schedule that cuts into profits and steals capital that could be used for growth. Suppliers do not want to make loans and will charge for their inconvenience.

In the New York metropolitan area, which includes northern New Jersey, southern New York and most of Connecticut, it is normal for an average household to use 1000 gal of heating oil over the course of a heating oil season. Western Pennsylvania, northern New York, and the rest of New England will use more. Points south of New Jersey will use less.

EXAMPLE OF HEDGING

The following example demonstrates one way to hedge against an adverse price movement. Fuelco, a home heating oil distributor, has pre-sold 42,000 gal of home heating oil to 42 households. Fuelco has agreed to charge these customers 99.9¢/gal and needs to make 40¢/gal profit. That means that Fuelco cannot pay more than 59.90¢/gal when it buys this fuel for the winter. Its *net* price under the rack cannot exceed 59.90¢/gal, or the company will lose money, customers, or both.

There are several different ways Fuelco can accomplish this objective. It can use futures, options, wet barrel programs, swaps, or its own storage. Each of these tools are examined in more detail later in the book, but the bottom-line for now is that Fuelco needs to use one or more of these programs. This is necessary in order for Fuelco to protect itself and its customers against the possibility that prices may move higher by the time the product is actually needed.

If prices do move higher and Fuelco has done nothing to protect itself, it will end up with a smaller profit margin. On the other hand, if prices move lower and Fuelco has done nothing, its profits would be higher. This is a trap into which many distributors fall. After living through a winter during which prices have dropped, many distributors have been tempted to believe that it may be better to leave prices unhedged. In retrospect, it has been at times, but there are ways to enjoy the best of both worlds. It is always dangerous to apply last season's lesson to the season ahead. (An explanation of how to achieve the best of *upside protection* and *downside potential* through the use of options or through wet barrel capped-price programs are given in later chapters.)

It is assumed that Fuelco is not interested in leaving open the possibility of lower prices and is interested only in locking in its 40¢/gal profit margin. It has two risks that need to be addressed to make certain it realizes its intended margin. The first of these risks is straight up-and-down *price risk*, or *directional risk*, where Fuelco needs to protect itself against prices moving higher. The second risk is *basis risk*, also called *differential risk*.

Racks are physical hose mechanisms that are lowered over tank trucks to fill them with heating oil or diesel fuel. These trucks then deliver the fuel to consumers or to gas stations that sell diesel.

Directional risk is the simple risk that prices may move higher by the time the product is needed. For a heating oil distributor, that risk is most closely associated with winter, when cold temperatures increase demand for heating oil. Prices sometimes drop during winter, if temperatures are normal or comparatively mild. Sometimes, though, especially severe temperatures can lead to skyrocketing prices. Political considerations can also make prices jump. Protecting against directional risk is the rather ordinary task of protecting against circumstances that can occur suddenly and that can make prices move higher.

Basis is the difference between the price at a central location (New York Harbor, or the NYMEX price is the one used most widely in refined products), and the price someone pays under the rack. If the rack Fuelco likes to use most is fairly near New York Harbor, it is unlikely to experience wide fluctuations in the basis between the two. Fuelco will frequently pay only 1.50¢/gal to 3.00¢/gal above the Harbor or NYMEX price. During the spring, summer, or fall, there will not be much change in this. Maybe, during a period of ample spot supplies, Fuelco will actually pay less under the rack than the spot NYMEX futures price.

It is equally possible that Fuelco will pay more during a tight period. The prompt premium for heating oil in New York Harbor reached 20¢/gal more than the NYMEX in December 1989. Prompt material fetched as much as a $1/gal over the NYMEX by January 2000.

There is almost always a surcharge collected by wholesalers to distributors filling their trucks under the rack. During the warmer six to eight months of the year, it is extremely uncommon to see the rack price deviate much from futures, except by the amount of the surcharge. This is generally 3¢/gal. In January 2000, the basis, or differential, was as much as $1/gal more than the NYMEX when supplies were impossible to get.

Fuelco's operating framework

- Fuelco buys under rack (which is 3¢/gal over NYMEX price) and adds 40¢/gal

- NYMEX Price + Rack + Basis = Fuelco's cost

- Fuelco adds its 40¢ margin to arrive at a per-gallon price for its customers

- 56.90¢ + 3.00¢ + 0.00¢ + 40.00¢ = 99.90¢

Fuelco's operating risks

- NYMEX price + rack + basis costs + 40¢

- The wholesaler's 3¢/gal markup and Fuelco's 40¢/gal margin are constants.

- The NYMEX price represents an element of risk in Fuelco's operating budget.

That does *not* imply that basis is a risk that can be ignored. It should not be ignored, even though it is only a major problem once every three years or so. Like fire insurance on a house, basis protection is not needed often. However, when it is needed, it is needed badly.

One classic example of this occurred in December 1989, following a November that was far from genuinely cold. Distributors sat down to their Thanksgiving dinners without realizing that they were about to need basis protection. They still did not know until the basis blew out in January 2000. No one knew about Mrs. O'Leary's cow in Chicago, either, until it was too late. Because one can never know these things in advance, it makes sense to have annual protection against the risk.

December 1989 was the coldest December in the 20th century. Demand for heating oil on the East Coast skyrocketed, and there were spot outages at various terminals in the northeastern states. The basis went wildly out of control. At their highest point, rack prices were a full 18¢/gal more than futures on the NYMEX. Traditional hedging methods were suddenly found lacking. Those who held futures or options found themselves trying to get their hands on the wet oil needed to fill their customers' empty tanks. The paper instruments just did not cover against the additional basis risk. The result was that heating oil distributors had to pay the difference out of their pockets. A valuable lesson emerged from that mess: no number of futures or options can fill a single boiler.

Wet barrels are the actual, physical barrels that are used in heating a home or running a vehicle. They are called wet because they are oil as a liquid. Traders use the term to differentiate physical barrels from paper barrels like futures or options.

One of the biggest problems facing rail companies each winter is the potential need for kerosene to cut the diesel that many engines need to operate. When temperatures drop below a certain point, kerosene must be added to diesel to maintain its pour point, or the temperature at which it will continue to pour. If temperatures drop far enough, rail operators will need large amounts of kerosene to keep their equipment moving. Companies that have ignored this need, especially in the Northern Plains, Great Lakes, and Northeast regions of the United States, have paid dearly for kerosene.

Kerosene is based on the underlying price of distillate, but it, too, can be subject to violent spikes above futures prices in cold months. It is not just heating oil distributors or end-users that need to be hedged against especially cold winters. Northern diesel users need to be aware of their kerosene needs when it gets bitterly cold, as well. They can benefit from hedging against sudden leaps in the spot price of kerosene by engaging their suppliers ahead of time to lock in their wet barrel needs.

This is far from being an indictment of futures and options as useful financial instruments. On the contrary, we believe that futures and options have valuable functions in hedging. It is just that one should be cautious about using them alone in December and January. In two out of three years, they may work perfectly. But there is always a chance that they will not completely mirror the change in cash prices under the rack. And that is what needs to be hedged for in the dead of winter. Months like October and March can be adequately covered by almost any instrument, including futures and options, while November and February only need partial protection with wet barrels. A realistic assessment of the risks suggests that they are not that great outside of December and January.

In order to protect against basis risk, one needs to contract with a supplier or with a broker specializing in wet barrel protection plans. It is wise to use a supplier familiar with the risks as well as the methods currently being used to contain them. A prudent buyer also would investigate how well its customers were supplied with actual, wet material in January 2000, and again in the winters of 2002–03 and 2003–04. If the supplier was able to provide the physical barrels then, it will probably be able to deliver the goods again.

The whole key to proper hedging for the winter heating season starts with an honest appraisal of risks. The biggest and most damaging potential risk is that a distributor may not be able to get actual wet barrels. There are certain times when different hedging tools are preferable to others. It is important to realize this before launching into any in-depth discussion of what may be involved in each one. The next few chapters will explain when one should use wet barrels or storage or inventory financing, and when to use paper instruments or financial methods to hedge the relevant risks.

Different Kinds of Risk

In the earlier energy markets, there were only two kinds of risk that distributors and wholesalers needed to be concerned about. The first of these was price or directional risk, and the second was basis or differential risk.

PRICE OR DIRECTIONAL RISK

Price risk is movement on the NYMEX. Perhaps a company has promised a large number of customers that they would be delivered oil during the winter at a specified price. In doing so, the company runs the risk that the underlying product may increase in price before it actually gains possession of the product or has established its price. This unprotected position can cause sleep deprivation brought on by a fear of prices moving in a specific direction.

The winter of 1996–97 was a case in point. Futures prices kept moving higher, with minor dips and valleys, from late summer of 1995 to autumn of 1996. Distributors never got the chance to lock in prices on any good dips, simply because prices never dropped significantly. Prices remained high because inventories of distillate were the lowest on record from the start of 1996 through the end of summer and early autumn. The winter of 1996–97 provides a good example of price risk, because prices were higher everywhere. Prices were not only higher than they had been earlier in the year, they also were higher than they had been in preceding years.

The winter of 2002–03 serves as another example of intense price risk and volatility. In mid-March 2002, nearby heating oil prices were less than 65¢/gal. January heating oil closed at 68¢/gal on March 11, 2002. Distributors working with 45¢/gal profit margins might have been enticed into offering budget customers a winter rate of $1.13/gal. However, by December 27, 2002, January heating oil prices had climbed to nearly 91¢/gal. In their peak demand month, unhedged distributors would have been out 23¢/gal (out of a 45¢/gal working margin).

The situation continued to deteriorate. On February 28, 2003, March heating oil reached $1.31/gal, up from less than 67¢/gal on March 11, 2002. March is often a stronger-than-expected month for demand, and in 2003, temperature readings in the Northeast were running a steady 10% colder than normal. Distributors who had not prepared (by hedging) found themselves in the worst of all worlds—forced to pay a higher price on a larger number of gallons than they had expected. In many cases, they simply had to raise prices and cut into their operating margins.

Unhedged distributors who promised budget customers a fixed price of 99¢/gal were often forced to absorb the difference as a loss. Many distributors lost money on budget accounts in those years, which forced them to be just a fraction less competitive with new customers in order to make it up. They had to forego their anticipated profit margins on their budget accounts.

Airlines were also hurt. In addition to the loss of travel volume in the aftermath of the terrorist attacks on September 11, 2001, airlines experienced dreadful fuel price volatility in 2002 and 2003. Jet fuel in New York Harbor could have been purchased for less than 64.5¢/gal on March 11, 2002. On February 28, 2003, the same product cost more than $1.30/gal. In less than a year, airlines saw fuel costs double. And this came in a year that was among the most competitive as individual airlines struggled to maintain market share in an industry that had lost travel volume. Estimates of the lost volume a year after September 11, 2001 were between 8% and 11%. The doubling of fuel costs during this period of low travel volume was especially devastating.

BASIS OR DIFFERENTIAL RISK

Futures and options are useful for protecting against the broad price risk of a strongly moving market. These risks can wipe out a balance sheet. But for end-users or distributors one level above them in the supply chain, futures and options are only able to carry part of the risk. One of the most important risks for end-users and distributors is *basis risk*, because sometimes they need the product right away.

Basis risk is the name for two different types of risk against futures—one based on time and one based on location. *Basis* is the premium or discount against futures for a specific time or location.

The difference between Boston Harbor and NYMEX prices is a location basis risk. The NYMEX price is for New York Harbor delivery. Prices in other locations may be above or below the New York Harbor price based on local supply and demand. The price between prompt barrels today and *any month* barrels two or three weeks away is a time basis risk. Buyers who need product today are often willing to pay more to get it delivered today. This is particularly true during periods of unexpected supply constraints or exceptionally heavy demand.

Any trader who hedges anticipated needs in Boston by using a New York futures contract can be seriously compromised when picking up the wet material under the rack. A large snowstorm may affect Boston more profoundly than New York and could make it more difficult for supplies to arrive in Boston. As a result, wet barrels in Boston may command a premium over New York.

NYMEX prices always reflect the price in New York. Boston prices are traded as a premium or discount to New York, based on the unique characteristics of Boston's needs at a given moment. NYMEX contracts are the only heavily traded energy futures contracts in the United States.

Time basis risk is the differential in price between immediate delivery and delivery at some point in the same month. Anyone who needs a barrel right away, such as today, can be compromised even when hedged by futures. With NYMEX futures contracts, barrels can be delivered any time during a full month, at the seller's discretion.

For a buyer who needs physical barrels immediately, especially earlier in the month, futures delivery at "any time" during the month may not be soon enough. This may necessitate the purchase of prompt, or immediate delivery, of material. The resulting prices can sometimes be higher than an *any month barrel* (or mmBtu) to be delivered at the seller's discretion. This is a time basis risk.

This risk happens most frequently with heating oil prices in winter. The differential can widen out against local rack prices, playing leapfrog with futures and easily winning before the game is over. The differential leaps ahead of futures, leaving the distributor to try to explain this phenomenon to home owners. In order to fill homeowners' tanks, the distributor winds up paying a hefty premium for prompt material. The end result is that the heating oil distributor ends up eating another goodwill effort.

Basis differentials trade as their own separate market. This is true for both time and location. Temperature differences, refinery problems, frozen waterways, economic disparities, and a number of other factors can raise or lower the basis for any given location and/or time against the NYMEX any month delivery.

Because both diesel and jet fuel prices are calculated against heating oil futures, one will just as frequently see huge price differentials for both products in locations outside of New York Harbor. When it is very cold in New York, prices for heating oil futures will advance. Prompt material prices in the Northeast are also likely to command a premium. Midwest or Southern diesel prices, however, are likely to be at a discount. They are not using the distillate to stay warm; they are using it for transportation purposes. And jet fuel prices will differ markedly from one part of the nation to another. In the Northeast, jet fuel/kerosene is used to cut heating oil to make sure that it can still pour in freezing temperatures. Elsewhere, it is used only for transportation or other needs.

In the early years of hedging in heating oil, traders left out basis risk and were hurt. Without prompt, wet protection, there is a risk of running out of heating oil if it gets cold. In the summer, gasoline resellers and marketers can run into the same problem. If driving demand is heavier than expected, because the economy strengthens or the weather is especially conducive to travel, there can be sudden increases in demand. Add in a refinery problem or unscheduled maintenance, and gasoline resellers or marketers can be crunched by both supply and demand. It is essentially the same risk experienced by heating oil end-users and distributors in winter.

Financial instruments are precisely that: *financial* instruments. They do not take care of supplying physical product. For that, in a pinch, one must go to local rack resellers, which can provide fixed-price and capped-price wet material. It is good to have these in place.

Even integrated refiners keep their upstream and downstream units as separate and distinct profit centers. A refinery owned by any of the large household names will pay its parent company the market price for crude oil, regardless of the price at which the parent company actually purchased it. The refining end of the company is expected to generate profits of its own.

In those situations in which a city or region is supplied by a fixed number of refineries, which is the case almost everywhere, the loss of one regional refining unit can dramatically alter the locational basis or differential there. This is especially true in California, where each part of the state is dependent upon a certain group of refineries. Imports can make up for lost supply, but they take time to arrive. That can inject time basis risk into the equation as well.

AVAILABILITY OR SUPPLY RISK

It had happened before, but January 2000 was one of the most notable examples we have seen of terminals running dry. Throughout New England and in New York, there were entire cities and surrounding suburbs that ran out of heating oil for a day or more. The unthinkable had occurred: the triage of a region in the face of cold temperatures. Prior to that, terminals had only run dry when temperatures had been severely cold. In January 2000, all it took was a cold snap in the middle of the *warmest* winter in 105 years to cause disruptions and spot outages. Spot outages have occurred since then, but largely as the result of bitterly cold temperature readings.

There also have been shortages of gasoline and jet fuel. In the case of the latter, cold weather was frequently the culprit, as well. Spot outages in gasoline have occurred because of the so-called Balkanization of its specifications. The United States has been moving towards an increasingly number of gasoline specifications since the phase-out of leaded gasoline in the mid-1980s. Most recently, there has been a move toward the use of reformulated gasoline in late spring and through the summer. But, it is not used everywhere in the country. It is mandated for use in some cities, counties, and states, leaving suppliers with the unenviable task of producing more than one kind of gasoline at the same time, often within the same supply region.

New York and Connecticut outlawed the use of MTBE (methyl tertiary butyl ether) as an additive to increase octane levels at the start of 2004. Some gasoline stations in those states have refused to sell anything other than unleaded regular because of this ban. The varying specifications for locations and seasons during the year have made it more difficult for refiners to react to regional disparities in supply and demand. The patchwork of specifications is what is meant by Balkanization of the product.

The *just-in-time* inventory system has been embraced in the oil industry and does not seem likely to lose favor any time soon. It has saved oil companies and their vociferous shareholders billions of dollars. But, in some years, at some locations, it just has not been in time. How often can one run out of gasoline or heating oil and maintain a viable customer base? Resellers and distributors *absolutely should not run out of supplies*. When they do, it pretty much defeats the whole purpose of their businesses.

As a result, distributors and resellers should move as far as they can as fast as they can from heavy dependence on one source. For December and January, and sometimes February and March, it is vital to have at least two heating oil or jet fuel suppliers. Three suppliers would be even better, and a terminal or tank farm in the backyard would be best of all, strictly from a supply point of view, of course. The same is true for gasoline retailers during the period from mid-May through Labor Day.

Supply risk is a relatively new form of risk that only became apparent in the late winter and early spring of 1996. And it is a factor in this market that is not going to improve without a concerted effort by the industry, and potentially even consumers, to rebuild an infrastructure that has been reduced and stretched past its initial objectives. The purpose of storage facilities is to prevent the consumer from running out of fuel. If the existing infrastructure cannot assure that supply, then it no longer meets its original goals. Over the past 20 years, as much as 60% of secondary storage facilities has been ripped out of the ground, for either financial or environmental reasons.

Primary storage facilities are those owned by refiners. Secondary storage facilities are those owned by distributors, resellers, and retailers. Tertiary storage is the heating oil tank in a basement or the gasoline tank in a car.

Between the ripping up of terminal space and the abdication of the refiners as holders of last resort, the cushion, especially for heating oil in the Northeast, is not what it once was. The people of Northeast region made it through the winter of 1996–97 because the weather turned warm at the crucial moment. In January 2000, it turned cold at the critical moment, resulting in a supply crisis. Distributors who were unhedged lost customers in droves to those who planned ahead. Some distributors were forced both to charge higher prices and to cut into their own margins. Customers were angry, and margins were low.

In order to avoid supply and price problems, distributors and retailers are encouraged to take as many steps as possible as outlined below. This applies to end-users and to wholesalers, as well.

Steps distributors or retailers can take

- The number of suppliers and terminals from which material can be lifted should be increased. Contracts from suppliers should require compensation for the full extent of any damages suffered through inability to provide contracted supplies.

- Storage facilities should be rebuilt or expanded. There are several ways of getting these facilities to pay for themselves, and they are worth the added comfort of being able to deliver wet oil when others cannot. The best strategy for beating the unreliability of suppliers is to build an infrastructure that provides a supply cushion. Without storage, a company is about as secure these days as a baby boomer planning to live on Social Security.

- Expansion of customer storage should be encouraged. The potential savings should be communicated to customers who fill up (with heating oil) in the spring or in July and do not need to buy more fuel again until winter is in its prime. That would involve an upgrade from a 275- to a 550-gal tank. In the case of an 1100-gal tank, most consumers south of New Haven can make it through 90% or more of the winter on fuel they can buy when it is cheaper.

  It may take a few years for the customers to recoup their investment, but it is worthwhile. It will be a capital improvement that will increase a house's value in the next decade when the big refinery crunch comes. There has not been a new grassroots refinery built in the continental United States since the early 1970s. Every refining unit in the United States is now critical to the area it supplies.

- Local, regional, and state associations should be encouraged to lobby for tax-credits for distributors, resellers, and consumers who do add heating oil, gasoline, or diesel storage. This storage would need to meet all the existing (and planned) EPA guidelines.

VOLUME RISK

Volume risk is most frequently associated with heating oil or jet fuel and can be caused by abnormally warm or cold winters. A business may have prepared for its customers to use an average of 1000 gal of heating oil but then may find them using significantly more in a very cold year. One way to protect against the possibility of needing greater supply is through buying *call options*. If temperatures are normal, they can be allowed to expire unused. If one needs 5%–10% more, they can provide a cushion. This applies to any end-user or reseller planning ahead.

The same is true for warm winters, when the risk is on the downside, through the use of *put options*. If one does not need the additional material, puts can protect against a weakening market with excess gallons looking for a home.

In most cases, it is better to overprepare for a winter by owning wet heating oil or jet fuel barrels (or having them contracted) and to protect against low prices with put options. The prompt market can go into a mild discount in a warm year, but it can advance to a catastrophic premium in a cold year. This preferred method is less expensive than buying both puts and calls. It also protects against basis risk. If 1100 gal of wet material are locked in, and a customer uses only 900 gal, puts will protect the company fairly well.

As noted, in those years with mild winters, when customers use less heating oil and prices are weak, the basis is unlikely to fall dramatically against futures. Those holding extra barrels can deliver them at their discretion during the month (when using futures or options on futures), because delivery is at the seller's discretion. As a result, prompt discounts in the cash market tend to drag futures prices lower with them.

During exceptionally cold winters, a customer might burn 1200 gal, and prices will likely advance. Under such conditions in the past, the prompt premium for an immediate or prompt barrel has climbed to as much as $1 more than the futures. Theoretically, that differential could be even greater. It is better to use a financial instrument to protect against downside exposure (lower prices) than against upside exposure.

It is important to consider all of the different risks facing one's business. The first is price risk or the market trend. The second is basis risk, the difference between an any month futures contract and the time and location at which material is needed. The third is availability risk, to make sure sources are lined up. And the fourth is volume risk, most generally associated with extreme temperature deviations. These tools can never fully or perfectly hedge against every potentiality. However, by planning ahead, businesses can anticipate the different risks that can be encountered and implement the appropriate supply strategies to ensure profitability.

SECTION II

THE SOLUTIONS: DIFFERENT METHODS OF HEDGING

There are many different ways of protecting against price risk. All of them have their individual merits. In the volatile world of oil prices today, it is not wise to deny oneself any of the benefits of these various methods of protection against adverse price movements. An integrated approach seems to be the best course, and wet barrel programs, futures, options, inventory financing, and one's own storage are all integral parts of this approach.

The discussion will begin with wet barrel programs.

Wet Barrel Programs

The following outlines the advantages of wet barrel programs, and it highlights the specific attributes that are important. Any wet barrel program should include the following:

- It must provide wet gallons. Some companies offer programs that in reality are little more than financial accounting tools. If the program is not going to provide the actual oil, it is cheaper to use futures or options.

- The supply *must* be available at a terminal that is accessible. Customers will need oil during inconvenient times, such as during blizzards, icy conditions, and nor'easters. If a supply terminal is out of the way, the company should have access to another terminal that is not. If one location is better than another, it should be made the primary terminal. The last thing any company wants is to add in long-haul costs to the other costs of doing business, especially if a terminal is difficult to get to during inclement weather. There are winters with so much snowfall that delivery times can be increased dramatically. If drivers take half a day to pick up product, then they are able to make only a few deliveries. This does not make sound financial sense.

• A wet barrel program should provide *firm insurance* against higher prices, but it should also be capable of protecting the company against downside moves as well. In today's world of cutthroat competition, a company cannot afford to let an unhedged competitor reap the benefits or the goodwill of lower prices. The best programs are those that are reasonably priced, offer a cap above which the price will not increase, but also allow the benefits of lower prices. This can also be accomplished with a fixed price program that is protected against downside exposure by put options or an equivalent instrument. These are often called *caps*.

• Any program should be reasonably priced. If a company is forced to give too much to its supplier, it can be put in an uncompetitive position relative to the competition. It is reasonable to pay a premium of 2¢–3¢/gal for a fixed price plan, and up to 4¢–5¢/gal for an *at-the-money* cap with downside protection during a normal year. During especially volatile periods, such as the one the United States has faced early in this century, these costs can get significantly higher.

These costs must be added to the company's cost of doing business before adding in the company's margin. This is very important, because a business cannot simply absorb these costs and continue to operate profitably. It is important to be aware of what is being offered as well as be prepared to shop around and compare programs, since new programs come out every year. One should never assume that a specific supplier will have the best prices just because it has had them in the past.

- Any program used must be priced in *local terms*. It is ineffective to have prices protected against moves in New York Harbor if the product is being sold in Norfolk, Virginia, or Madison, Wisconsin. The price contracted must represent wet barrels under the rack in one's neighborhood. It is vital to read the fine print and be certain of it.

One huge oil refiner offered a program that paid the difference between the New York Harbor price on the day one signed up and the New York Harbor price on the day the price was set or activated. No wet barrels were moved, and there was no location basis protection, so it was little more than a glorified futures program. The fact that it was offered by a large and well-known refiner did not ensure anything. These basic ingredients should be expected to be a part of any plan offered by an oil company.

It cannot be taken for granted that the desired protection is provided just because the supplier should have anticipated the need for it. This can be an expensive lesson. One must know at *what price* product will be lifted, *where* it will be lifted, and *when* it can be lifted. It is important to discover if January barrels can be lifted at one's own discretion or at the supplier's discretion. When oil is needed, oil is needed.

WHY THESE POINTS ARE SO IMPORTANT

Some heating oil distributors still feel that it is best to leave well enough alone. All these fancy programs just run up the costs of doing business. Some years, it may seem that the best way to go has been to pick up material under the rack and to pay as product was needed. That has certainly been true during warm winters in which the seasonal tendency for weakness is active.

The 10 warmest years on record all occurred in the years between 1980 and 2002. Prices developed a seasonal pattern of dropping from early autumn through late winter. The winters of 2002–03 and 2003–04 were significantly colder than many were used to.

However, that approach was disastrous in the winters 1989–90, 1990–91, 1999–2000, 2002–03, and in 2003–04. As mentioned in an earlier chapter, in December 1989, temperatures plunged to their coldest levels in 100 years. Heating oil prices skyrocketed. Basis risk, the difference between a cash price at any point and the expiring month in NYMEX futures, flew wildly out of control. In locations near New York Harbor, the basis was distorted as much as 20¢/gal more than the NYMEX.

Even without record cold temperatures, in January 2000 the basis increased to as much as $1/gal for delivery in New York Harbor. If supply was needed inland, it cost more. In March 2003, New York Harbor gallons commanded more than 22¢/gal over futures, which were already more than $1/gal. These instances are no longer rare, indicating that hedging is more important than ever.

Dramatic movements in absolute (futures) price and basis can destroy a business or its customers' goodwill. This was demonstrated in 1990, when the basis remained under control, but prices still took off again. Saddam Hussein invaded Kuwait with a goal of controlling or destroying its oil production. By the time he was sent packing, hundreds of oil fires had turned Kuwait into a surrealistic version of hell on Earth, and heating oil prices had more than doubled.

After an especially warm winter with generous supplies in 1998–99, prices almost tripled in January 2000. When the United States decided, in 2003, to end Saddam Hussein's rule once and for all, prices increased by a factor of 2.4 in the 15 months leading up to the actual war.

Those who did not have wet protection in December 1989, January 2000, or in February or March 2003 were hurt badly by both higher prices and an extremely wide basis. Those who did not have either futures or wet protection in 1990 were hurt only by skyrocketing prices since there was not a wide basis differential. Those who were unhedged had only two options: either pass the higher costs along to consumers or absorb the higher costs themselves.

The end result is that in two of the four winters between 1989–90 and 1992–93, it would have been better not to hedge, but in the other two years it would have been disastrous *not* to hedge. Those who were not hedged at all during the four years alternated between feast and famine. They made more than expected in the soft years, but were unable to pass costs along fast enough during the years with rising prices. And their customers would not have had any peace of mind. To them, every bill was like an adventure in an expensive restaurant where one buys an item marked "seasonal price." The question really boils down to a simple matter of insurance. Can a company afford not to pay a small premium to avoid disaster in the years when prices are out of control?

Wet barrel programs need not be used all year long, or under every circumstance. But there are clearly situations that cry out for wet barrel protection. Paper instruments, like futures or options, can be used during other times of year. In some situations, long-term derivatives can make sense. These hedging vehicles are not mutually exclusive. They can be blended together in an almost infinite variety to fashion an individually tailored hedging program.

WHEN TO USE WET BARREL PROGRAMS

Heating oil and diesel

Wet barrel programs are absolutely necessary for the months of December and January, and depending upon the latitude, should be considered for any or all of the months between November and March.

The following minimum coverage levels are recommended between wet barrel programs and company-owned or leased storage.

- November........................5%–20%
- December........................25%–50%
- January............................25%–50%
- February..........................15%–35%
- March...............................5%–20%

It is important to remember that these are *minimum* coverage levels.

During the summer months, there is much less risk that the basis will blow out like it might during the colder months of the year. During that time, futures and options are perfectly acceptable tools to use to protect against sudden and adverse price movements.†

Distributors and end-users can use these financial tools, which are frequently cheaper than wet programs, to protect themselves. Prices are protected, and material can be picked up from normal sources at

†*Futures trading involves risk and can result in substantial losses. Neither this book nor its author makes or represents any offer to buy or sell futures, or options, or other financial instruments. This is for educational purposes only. One should consult a qualified broker before engaging in these activities and be aware of the risks involved with any and all transactions. Substantial losses are possible.*

numbers very close to where they would be expected. The risk of a huge move in basis is negligible after mid-spring and before mid-autumn.

Gasoline

While the differential or basis risk is lower with gasoline, problems are often seen in late spring or early summer. The use of wet barrel programs is recommended for gasoline users and resellers for the months of May, June, and July. Slightly smaller amounts may be judiciously employed for April and August and even September. These are all months during which there have been unexpected prices moves. During the last 20 years, May has been the month with the most dramatic moves. Wet barrel programs (for gasoline in late spring) are especially useful for gasoline users in California and in the Great Lakes region.

Recently there were decisions by 3 states to outlaw MTBE (New York, Connecticut, and California), and 14 other states are committed to banning its use. Consequently, gasoline resellers now need to be aware of potential problems in the region they service.

Hedging gasoline is much more difficult than other elements of the oil complex. Prices are more competitive, and the transaction costs of using caps or of buying calls against product in storage make many gasoline resellers avoid hedging approaches. These costs must be added to the purchase price.

It is important to consider the following when implementing these programs:

- If a company has only one or two terminals that it can currently use, it should be certain that it can get supply even if others cannot. Some wet programs can offer this peace of mind.

- A company definitely needs wet barrel protection if it has little or no storage of its own. Remember, the whole purpose of this exercise is to be certain of obtaining burnable oil products. Anything less gambles with the customers' warmth or mobility.

- A company that offers fixed or capped price programs to its customers needs to be certain that it can get product at a price that will still accommodate the desired profit margins. Because these can be affected by movements in both the absolute price and in the basis at one's specific location, wet deals should be used to guard against adverse price movement or supply availability problems.

It makes sense to invest in wet barrel protection in gasoline in the following situations:

- When there is a threat of political developments or specification changes that can lead to higher prices and/or possible shortages. The reseller must strive for protection against a repeat of the situations that led to gasoline lines in the 1970s. Recently, there were problems getting gasoline from Venezuela during the general strike of December 2002 to February 2003. Had that happened in May, it could have been much worse. Seasonal grade changes also sometimes result in shortages.

- When a company's terminal supplier (or branded supplier) has limited storage space in that region. In that case, it is wise to have backup arrangements with other suppliers in the area. This is especially true in California and in the Great Lakes region. Since no pipelines run laterally, any refinery problem in either of these locations can cause severe tightness and explosive surges in prices. Late spring is the period of greatest risk, while refineries are in turnaround just before any switch to summer grade specifications.

- When a company has severely limited storage facilities or limited access to lease facilities of its own. Ideally, a company should have its own supply cushion. If it does not, it might want to engage in some wet barrel protection.

- When a company has municipal or fixed/capped price customers. It is always best to be certain these folks will be supplied.

- When all the different grades needed in an area require split storage facilities. This is a problem that has become chronic in recent years and may get worse. If there is any chance that a company will run out of an important grade because of the sheer number of different grades required in its region, it is best to ensure its source of supply. Major cities in California and in the Great Lakes region run the greatest risk of running out of summer grade gasoline in late spring.

Branded resellers are often protected against these possibilities. If this is not the case, the company may need to enter into supply agreements of its own. As is the case with heating oil, there are seasonal risks that seem to affect certain geographical regions more profoundly than others. Understanding these local risks is crucial to surviving in the market.

Gasoline stations generally do not yet offer the fixed and capped price programs that heating oil distributors offer to their customers. It will start with municipal fleets, police cars, buses, and taxi fleets, but eventually it will be offered to consumers through debit cards. This will prove to be a very useful method of instilling customer loyalty. Until that time, companies can enter into wet barrel programs where prices or basis differentials can be set at any time, even a few days before delivery. Wet barrel programs not only protect against price or basis movement, they also can be used to assure availability.

Jet fuel and kerosene

Wet barrel programs for jet fuel and kerosene should be used in a number of situations:

- In the absence of time or location basis protection in the winter. Frequently, the basis can be locked in at a certain point in time, and the underlying price can be locked in at another time. In general, it is good to lock in the differential whenever it is less than 2¢/gal. This is especially desirable for heating oil distributors, railroads, and transportation companies operating in very cold climates. Kerosene is used to cut heating oil and diesel fuel to make it pour in especially cold weather. During very cold weather, kerosene can command premiums of 15¢–20¢/gal more than heating oil prices.

- When airlines are offering cut-rate fares all at the same time, especially in winter. Here again, it makes sense to lock in differentials whenever they are less than 2¢/gal (for as much of a year as is possible). Airlines with more than one hub should also try to lock in different locations against each other (when they are priced close together) to help mitigate regional variances in jet fuel prices. These can become extreme, especially during the colder six months of any year.

- When military aircraft are likely to be especially active, such as during times of political unrest. This does not always affect the United States, but it does need to be included in the risk assessment of airlines operating internationally.

• Because of the different types of jet fuel, it is critical to keep up with developments that may lead to supply disruptions. Kerosene is used in the winter in the northern states to bring down the pour point of diesel or heating oil, and it is also used for space heating in the winter. If there is a danger that it will be in short supply, it is best to just lock it in through supply agreements. These can also save money during periods of rising prices, times often affected by the same circumstances that push underlying distillate prices higher.

The price for jet fuel or kerosene is usually quoted in terms of a differential against heating oil futures or heating oil in New York Harbor.

Understanding Futures and Options

One of the most misunderstood and potentially frightening prospects for businesspeople is the topic of futures and options. Futures and options contracts move quickly, and they are leveraged heavily, so that relatively small moves can result in huge profits or losses. They are also both *zero-sum* markets, meaning that no fresh capital or equity is ever created or destroyed.

In the stock markets, capital or equity can be spontaneously generated or consumed by a move in the market. In stock markets, these gains or losses are unrealized, or exist only on paper. One needs to close out the position to realize the profit or loss. Everyone can have a good or winning day or a bad or losing day.

That is not the case in futures or options on commodities exchanges (in the case of energy futures, the NYMEX). At the end of every trading day, traders have paper gains or losses that are the result of any price movement that day. For every dollar made, there is a dollar lost in futures or options on futures. And somebody has to pay the losing dollar right away, the very next day.

In commodity futures, the gains or losses are real regardless of whether or not the open position is offset. If prices gave a contract position a profit for that day, the profits are deposited into a margin account and can be accessed right away. If a contract has a loss in an open position, the holder must make a deposit in the margin account to make good on the lost funds the very next morning. Any dollar gained on a winning position must physically come from a dollar lost on the opposite or losing position. Since there is one contract held long (or previously bought) for each contract held short (or previously sold), the total dollar value remains constant. Losers, in effect, pay the winners each and every morning on every open position held in the market. This gives a certain unforgiving quality to trading in futures.

Hedgers need to be aware that the futures positions they may use to offset a physical risk may need to be funded. If a railroad needs diesel three months down the road and buys futures to protect against higher prices, it will need to put up more money to hold the contract in the event that prices decline. When it comes time to actually buy the physical product, the railroad will pay less for it, but it will have had to have deposited funds as the market declined. The railroad will not pay more than the contracted price, but it will pay part of it ahead of taking delivery and then the rest at delivery time.

Conversely, if prices rise, the railroad will have money deposited into its account as the price goes up, but it will end up paying it out for the higher priced physical product when it comes to time to take physical delivery. The end result is that the railroad will pay the price it has contracted. But it may have to provide a cash outlay, or it may receive cash, on its futures holdings. These will be offset by physical profits or losses at the time of delivery.

Example

If a trader buys a heating oil or gasoline futures contract, it controls 1000 bbl, or 42,000 gal. *Margins*, or the deposit required to buy or sell a contract, are generally a small proportion of the total worth of the contract, sometimes less than 5%. A trader could put down $1000 to control one contract, and then see prices drop 2¢/gal that same day. In this case, the trader would have lost $840 of the $1000 margin put down. Each 1¢/gal move in heating oil or gasoline is the equivalent of $420, or 1% of 42,000 gal.

In contrast with the stock market, in futures one can just as easily sell something one does not own as buy something one does not want to own. Since long and short positions in the same month of the same commodity cancel each other out, it is strictly a financial transaction. If one sells something higher than one buys it, one makes money. It does not matter which is done first.

The following morning, the trader who bought this contract, which dropped 2¢/gal, would need to deposit an additional $840 with his broker. Had prices advanced 2¢/gal, the buyer would have had $840 deposited into his brokerage account the following morning. Money changes hands rapidly, but one will never pay more than the total value of the contract price that is locked in. Any money lost on a day-to-day basis is balanced by the lower cash purchase price if delivery is taken at the closing of the futures position.

If a futures contract is purchased at 80¢/gal for delivery three months into the future, and futures prices drop to 70¢/gal, there would be a 10¢/gal additional margin requirements ($4200) as the price slipped. When delivery is taken at the lower cash price of 70¢/gal, the 10¢ margin cost will be recovered. The end result is that the purchaser pays a 10¢/gal margin call as the price drops and pays 70¢/gal cash for the product, which is the equivalent of the 80¢/gal that was budgeted for three months earlier.

There are huge risks associated with futures and options. Futures prices will balance outstanding cash positions, but the risks need to be understood. If a margin call cannot be met, the open futures position (used to protect against a move in cash) will be closed out by the broker-age house. If a contract holder does not meet a margin call, he will still be responsible for any financial losses, but his broker has the right to close out his positions. In that case, the protection against a move in the cash market is lost.

FUTURES CONTRACTS

A futures contract is an obligation to either take or make delivery of a specified number of units (gallons, barrels, bushels, ounces, etc.) of a given commodity at a specified time and location in the future. In the case of oil, each contract is 1000 bbl, or 42,000 gal. Natural gas contracts call for delivery of 1 mmBtu.

There are 42 gal/bbl. However, there are no actual, physical barrels any more. This is used strictly as a unit of measurement.

Each contract has very detailed specifications, including factors like pour point, sulfur content, Reid vapor pressure, viscosity, and cloud fac-tors. These contracts are based on Colonial Pipeline specifications for refined products: Cushing, Oklahoma for crude oil and the Henry Hub in Louisiana for natural gas. As such, they are *fungible*, or interchangeable with the underlying physical commodities used as the standard in the world of cash transactions.

Fungibility is the ability to exchange one contract for another. Every contract is interchangeable, which is one of the preexisting conditions needed for a successful futures contract. Consequently, if someone buys a December heating oil contract, the sale of a December heating oil contract will cancel it out. It is critical that one understands that in futures, one can sell a contract one does not yet own. One is *short* the contract when one sells something one does not have. Since there are times when one does not plan on honoring the contract with delivery of the commodity, one will need to offset it. In this case, this can be accomplished by buying a contract against it.

The existing contract is canceled out by an opposite transaction. If one sells it, one needs to buy to offset it. If one buys it, and gets *long*, one needs to sell to offset it. In futures, there is no sort of distinction between which is done first, in terms of buying or selling contracts. A contract is only an obligation to make or take delivery of the specified commodity at a given point in time. Because futures are based on underlying cash instruments, those traders who do want to make or take delivery of the physical commodity can do so.

Longs and shorts cancel each other out, like matter and anti-matter. If a person is holding one contract long, or has previously purchased a contract, he can offset it by selling one contract. The previously held long position is canceled out by the new sale, or short position.

Margin calls

In any discussion of futures, it is important to be aware that there are margin calls that need to be made immediately. If a futures contract is bought in heating oil at 55¢/gal, and it drops to 54¢/gal, a deposit of $420 is required per contract of 42,000 gal with the NYMEX. If prices advance to 56¢/gal, that money could be withdrawn, at least for the day. If prices return to 55¢/gal the following day, the NYMEX would ask for it back. When money is lost on a futures position, NYMEX will contact its clearinghouse, which in turn calls upon the brokerage house through which the position was originated. The position holder will then be contacted by his broker to make any required margin calls.

A 1¢ move on 42,000 gal, or one contract of 1000 bbl, is equal to a gain or loss of $420.00.

Moves can be very abrupt in futures. Because futures are *obligations,* position holders can be liable for a large amount of margin money at very short notice if the market goes against them. This has an unsettling effect on most distributors and is one of a few major reasons why many prefer wet barrel programs. There are advantages to futures, if they are understood and managed properly.

Futures offer greater liquidity, and they are the most inexpensive instrument to use in terms of transaction costs and the number of points lost while actually getting in and out of positions. If a company is in a position to take delivery, then they are far and away the best instruments to use, but not many companies have the ability to dock a barge out back. If taking delivery is a possibility, though, it is something that should be looked into. They are an inexpensive way to get a rein on elective or optionally hedged barrels.

Example of long hedge

The One-Day Delivery Two-Oil Company (Odd Two Co.) has initiated a fixed-price plan for its budget customers and launches it in March of the preceding spring for winter gallons. It is figuring on a normal usage of 1000 gal per home during the months from October through March. Approximately 500 customers have signed up for the program by the end of March, and Odd Two decides to hedge 12 contracts on the NYMEX. It simultaneously locks in a differential of 3¢/gal under the rack with its normal supplier, thereby protecting itself against any adverse price moves at its local level.

Odd Two likes to make 40¢/gal to take care of overheads and to provide an operating profit of 10¢–20¢/gal, after overheads (rent and salaries) and depending upon other factors. (These could include road conditions, maintenance needs, prevailing interest rates, and other miscellaneous factors.) After adding in the 3.00¢/gal under the rack and the 40.00¢/gal operating margin, Odd Two comes up with a fixed price of 99.9¢/gal.

The company and its customers are comfortable with that price. They would prefer not to pay more than $1/gal. At the end of March, Odd Two buys a number of contracts over six months to protect against higher winter prices. It decides to buy 12 contracts of 42,000 gal each, and it spreads them out over the six colder months of the year (see Table 5–1). It purchases the following contracts in the following numbers:

Table 5–1: Odd Two's contracts

Month	# of Contracts	Price (¢/gal)
October	1	56.00
November	2	56.50
December	3	57.25
January	3	57.50
February	2	57.25
March	1	56.00
Average Price	12 lots	56.98

Odd Two buys one contract each for October and March, two contracts apiece for November and February, and three contracts each for December and January. With 42,000 gal in each contract, Odd Two has locked in prices for 504,000 gal of heating oil. This is 4000 gal more than the number it expects those customers who already have signed up for its fixed price program to use. It views that as an acceptable match.

By buying these contracts and by locking in the basis or local differential against New York Harbor, Odd Two has locked in a price of 99.9¢/gal for its budget customers. If prices move higher for any reason, this 99.9¢/gal price still will be safe. The downside is that if prices break sharply, Odd Two still will need to charge 99.9¢/gal to these customers. Futures are an ironclad contractual obligation. In the same way that futures would be expected to move in tandem with cash prices if the market moves higher, so, too should they be expected to drop along with cash prices.

There is another potential pitfall here: margin calls. If prices were to fall 5¢/gal, Odd Two would have to ante up another $25,200 (12 contracts x $0.05 x 42,000 gal) to maintain its long positions. This might be difficult over the summer when revenues are generally lower for most distributors. Due to potential margin requirements, many companies prefer to go straight to their suppliers and work out wet barrel deals. Then they do not need to pay anything until they take actual possession of the physical barrels.

The greatest advantage of futures is their liquidity. If something unforeseen happens, futures positions can be exited and hedged again later at a lower price. Generally, it is more difficult to get out of a wet barrel program. The purchase of put options or the sale of futures against the wet barrels could well serve the same purpose. One should have the ability to trade futures if needed but also should be prepared to do so in a disciplined and careful manner.

Example of a short hedge

Romeo Refining has a refinery in Joliet, Illinois, which has decided to produce an extra 10,000 bbl of distillate in each month during the fourth quarter because of high refining margins. It already has locked in its regular production but now wants to hedge the additional production against any potentially sharp drop in refined products prices.

Perhaps its best deal would be to lock in the lifting prices for some of its regular diesel customers. That way, it can also protect the basis between Joliet and New York. It would include the basis and a terminal charge in its quoted prices. Otherwise, it would need to either absorb any potential drops in fourth quarter cash prices against futures or work with a bank on getting swaps to protect against that risk.

To protect against price (or directional) risk, though, it would need only to sell futures: 10 contracts each in October, November, and December. By selling futures outright, Romeo foregoes the possibility of making more if prices advance. However, it also does not need to worry about the possibility of a collapse that could have a negative impact on its refining margins. If refined products drop by a combined average of 10¢ during the fourth quarter, Romeo will sell those products for 10¢ less than anticipated in the cash markets. However, it will make up for that with 10¢ gains in its short futures positions.

If Romeo had wanted to be able to capitalize on any potential upside in prices without leaving itself open to the chance of losing money if prices fell, it could have used options. They can be more expensive to use, but they do allow a company or an individual to keep the upside potential open without worrying about the downside, or vice versa.

People who trade futures

There are two basic classes of traders in any futures or options contract: *commercial traders* and *speculators*.

If one is holding product, one is long.

If one needs to buy product, one is short.

The first group is trying to lay off or *hedge* existing risks (those risks that come naturally just by being in a business). For any business, the risk is one of inventory needed or inventory held on hand. A person who needs to buy material is short of it and needs to balance or hedge the risk against the potentiality of rising prices by placing a *long hedge*. A person who is holding excess product is long on it and needs to place a *short hedge* in order to protect against lower prices and inventory losses.

A long hedge secures product against a naturally short position and allows companies that will need product to enter into a contract to get it. They will need the product at a future time, which makes them short of it. By placing a long hedge, they are securing that future need at a price that will protect their operating margins. Long hedges are especially helpful for end-users.

A short hedge offsets a naturally long position by protecting inventories against a decline in prices. Furthermore, it allows a company to lock in a future selling price. That is especially desirable for a producer. In both cases, operating profits and debt-servicing needs can be nailed down.

Crude oil example

In the case of crude oil producers, they will need to borrow money from a bank to purchase the lease, explore, and drill. Their bankers are not going to be thrilled with the idea of selling the crude produced on the property at some as-yet-undetermined price. Crude oil prices have traded as low as $9.75/bbl and as high as $41.15/bbl. The bank is not keen on taking the risk that a company will sell its crude oil at $10/bbl in exchange for the possibility that it might be able to sell its crude at $40/bbl. The bank would prefer to know that the company has contracts in place to sell its crude at $25/bbl. This is particularly true in the United States, where the typical cost of production is $12–$18/bbl. Each producing field has its own unique set of economic factors. The bankers (and shareholders) would prefer to know that each field is capable of producing at a profit.

The opposite case is true for an airline. Fuel costs are one of the largest variable costs it needs to know. Fuel costs are an airline's second largest expense, after payrolls. It is difficult to sell tickets profitably six months down the road without knowing what the variable fuel cost will be. Vacationers are not thrilled with the idea of buying a ticket if the price will be established at some later date. They need to budget for their vacations. At the same time, corporate shareholders are not going to be thrilled about having the holiday season completely booked if every flight yields a loss after fuel costs are added in.

Clearly, an airline has several other considerations, such as the perishability of seats, that must be taken into account. But, changing physical fuel costs can play havoc with the bottom-line. Knowing what those costs are going to be ahead of time can help an airline in its pricing decisions and can help it to remain profitable.

By hedging its fuel costs six months ahead, an airline can make rational pricing decisions. It can advertise its rates to attract travelers. Unfortunately, it must also consider its competition's rates, which are sometimes not made rationally. However, by limiting fuel costs ahead of time, it has something with which it can plan ahead. If it does not, it is playing roulette with shareholder equity every time it books a seat six months down the road.

For those who need to buy the product, and need to establish a long hedge, futures are likely to rise by an amount similar to the amount that cash prices will rise. So, if gasoline is needed in June, a futures contract purchased in March is likely to advance or decline by an amount equivalent to the price for cash. This takes us back to price risk or directional risk, which was discussed in the previous chapter.

There is a catch, though. If the hedger buys gasoline futures for June delivery in March by buying a futures contract, and prices then decline, that person will have to pay margin calls regularly. There will not be a corresponding cash influx from the declining price of the physical product actually needed in June. On the other hand, if prices rise, the advancing futures prices will add cash to the hedger's brokerage account, deposited by the individual who sold short. In that case the hedger will not need to meet any additional margin calls.

Cash and futures move together. However, only futures are settled up each and every morning. If the market moves higher, one will have compensating gains in futures to offset the higher price needed to pay for cash or physical barrels. If prices decline, though, one will lose in futures and will need to settle up each day until reaching the actual time the product is needed. Cash prices will fall as well, meaning that the money one pays to keep the futures position or contract will reflect the lower cash price eventually paid in the cash market.

Refined products example

A gasoline trader needs 42,000 gal, or one contract of May gasoline. She is afraid that prices will go higher. She is short—or does not yet have—gasoline for May. On March 5 she buys one gasoline contract at 70.00¢/gal as a long hedge against higher prices. Cash prices for May are trading at 70.50¢/gal on March 5, or at a premium of 0.5¢/gal to futures.

Through the rest of March and April, May gasoline futures prices advance 10¢/gal to 80.00¢/gal. Cash gasoline advances to 80.50¢/gal. This gasoline buyer pays for her cash gasoline and pays 10¢/gal more than she would have in March. But she makes 10¢/gal in futures, which has been deposited in her brokerage account. The end result is that the buyer pays the equivalent of 70.50¢/gal, the price she liked in March. The 10¢/gal gain in futures offsets the 10¢/gal loss in the cash market.

The catch outlined above occurs when both markets decline in March and April. For example, consider the case of the gasoline buyer at 70.00¢/gal when cash is at 70.50¢/gal. By the end of April, the futures price has fallen 10¢/gal to 60.00¢/gal. In May, the gasoline trader now buys physical gasoline 10¢/gal lower, at 60.50¢/gal. But she has had to pay 10¢/gal in margin calls, or $4200 to stay hedged with futures.

The end result is the same. The cash price is 10¢/gal lower than the 70.50¢/gal price desired in March (at 60.50¢/gal). But the futures contract has lost 10¢/gal. The trader still owns her gasoline at the originally desired 70.50¢/gal at the start of May because both cash and futures have fallen 10¢/gal. But she has had to come up with $4200 in margin calls to keep the hedge operational. The catch is strictly one of cash flow. By the end of the exercise, the dollars gained balance out the dollars lost.

As a point of reference, the 0.50¢/gal difference (between cash and futures) used throughout this example is the basis. In reality, this number will change. Right now, though, the concern is price or directional risk alone. In the case of this risk, what is important to grasp is that cash and futures prices will move in the same direction together. It is rare that the two will actually move perfectly together. However, with the large moves often seen in oil prices, it is likely to be in excess of a 90% correlation 90% of the time. Methods to protect against basis risk will be covered after a thorough explanation of the methods used to contain directional risk.

The role of the first group of traders in the market, hedgers, has now been outlined. They have or need physical product and need to protect against price moves that can dramatically alter their bottom-lines. In the case of the gasoline buyer above, she may have agreed to deliver gasoline in May to a fleet of automobiles or a marina at a set price above 70.50¢/gal (which was the cash price on March 5). If she needs a 12¢/gal profit or operating margin to stay in business, a prearranged selling price of 82.50¢/gal may have been agreed upon with the end-user on that date.

In the example above, when prices rose to 80.50¢/gal, if the buyer had not been hedged, she still would have had to sell at 82.50¢/gal to the automobile fleet or the marina. Her operating margin would have been cut to 2¢/gal in that first scenario (82.50¢/gal selling price minus 80.50¢/gal buying price). Granted, if it had not been hedged in the second scenario, where prices dropped to 60.50¢/gal, the profit margin would have ballooned to 22¢/gal. (This is calculated from taking the 82.50¢/gal selling price minus 60.50¢/gal buying price.) But the point of hedging is to remain in business, making operating (or normal profit) margins. If one is interested in letting the market hand out feast or famine, one may, in fact, prefer to be a member of the next group of traders: *speculators.*

Speculators are private individuals or pooled funds that wish to assume the risks that commercial traders wish to lay off. Speculators are frequently maligned, but they provide market liquidity and shoulder the risks that businesses want to hedge. For that task they are given the opportunity to make trading profits. Strictly speaking, commercial accounts are not interested in trading profits but are instead interested in protecting operating profits. That is the theory of it, although most large commercial end-users and producers actually have trading arms that also try to make trading profits.

There is a big difference, though, between large commercial players and large trading fund players, even though it may not be apparent in the actual trading ring or in the way that prices move on the screen. The commercial trader works for a company that got into futures to protect its operating income, while the fund trader works for a company designed to make money in futures and related instruments by trading. Commercials already have risks associated with owning or needing a commodity. Traders go out looking for that risk, and for the potential rewards shouldering that risk can provide.

How do speculators differ from someone gambling in Las Vegas? Futures traders are making or losing money because they have assumed *real risks*. They are willing to assume risks (and rewards) that businesses involved with commodities do not want. A gambler is making or losing money at Las Vegas based on risks that do not really exist. They have been invented solely to determine a winner or loser at the gaming tables. There is no real risk in Las Vegas until a player sits down at a blackjack table. The dealer does not deal to an empty table.

A heating oil distributor has price risks associated with first buying and then selling heating oil. Before a distributor buys product, the price needs to be determined. If it is not determined in advance, the price can be considerably higher than originally expected by the time it is needed. Once the material has been purchased and is held in storage, prices can fall. These are risks a heating oil distributor would rather not have. Someone needs to be found to assume them.

Speculators assume those risks. The risk on distillate prices exists in the real world with or without speculators. The risks in Las Vegas exist purely for the gamblers, or really for the house, and would not be there without them. There is not any risk at an empty casino (except for poorly connected managers). There is always a risk in the petroleum business. Speculators assume risks that already exist. Casinos create artificial risks that gamblers then willingly assume.

Businesses that start out with the risks associated with petroleum price movements have an inherent risk. When these businesses do not hedge, they are, in fact, speculating. They may never have gone out looking for risks or done anything consciously to create those risks. But they have them simply by being in business. Every company that has or needs any petroleum product has those inherent risks. Prices will move up and down. The only way to negate those risks and stop speculating is to hedge those risks by passing them on to someone who is consciously and deliberately willing to assume them.

Any company that uses or produces commodities, including crude oil, refined products, and natural gas, has a natural exposure to price volatility. Individuals who understand the benefits of hedging and have taken the time to fully educate themselves on available hedging methods will help to ensure a company's growth and viability. This understanding is essential to success in these businesses.

How prices are determined in futures

As they are in most markets, prices are agreed upon after a process of buyers making a bid and sellers making an offer. In commodities trading rings and pits, these bids and offers are made by *open outcry*, which allows any two traders to agree upon a trading price. Sellers must take the best (or highest) bid, and buyers must take the best (or lowest) offer. When a buyer and a seller agree on a price, it becomes a trade and is listed as the most recent price for that commodity.

If Bob, a trader, wants to buy crude oil, and Susan, another trader, wants to sell, this is what could happen: Bob could bid $31.00/bbl to buy crude. Susan could offer to sell at $31.10/bbl. That makes the market *even bid at 10*, dropping off the first numbers or the *handle*. Bob wants it pretty badly, so he raises his bid to $31.05/bbl. Susan is willing to get closer to trading, so she lowers her selling price, or offer, to $31.07/bbl. The market is now *5 bid at 7.* Bob is more eager, and he raises his bid to $31.06/bbl, by saying, "6 bid." Susan agrees to that price, and says, "Done." The price has traded at $31.06/bbl, which becomes the market's last or most recent price. That is the price that traders in offices will see on their screens.

OPTIONS ON FUTURES CONTRACTS

An *option* is *an opportunity without the obligation* to buy or sell a commodity at a given price (the *strike price*) in the future, in the event a different option is purchased. In many ways it is the inverse if an option is sold, when it becomes an *obligation without the opportunity* to buy or sell a commodity at a specified price. When one buys an option, one pays a premium. When one sells an option, one collects the premium. It is very similar to a life or fire insurance policy, and the premium is priced along parallel lines.

Volatility *is how quickly* **underlying** *prices are moving.* **Time value** *is the amount of time left on the option, and* **intrinsic value** *is when an option is* **in-the-money,** *or when it is already making money.*

Instead of actuarial tables, options sellers use a *black box* to compute a fair-value premium based upon volatility, time value, and intrinsic value. The sellers of options collect premiums in the hope that it will never become profitable for the buyer to *exercise* or use them. The buyer of an option is using it to establish a minimum selling price or a maximum buying price.

Understanding these instruments is important, and yet it requires a certain amount of reinforcement. What is an opportunity without the obligation? It is an opportunity to be long or short a given futures contract at a specific price (the strike price) but it is *not* an obligation to be long or short that contract, should one choose not to be. An option holder does not have to exercise the option and can let it expire instead. He pays the premium for this opportunity, and the premium is the most he is going to lose. If the market moves opposite to the direction that he was expecting or hoping for, he can let the option expire worthless rather than exercising it at a price that would lose more money than the premium.

There is another very valuable aspect of buying options that can help companies as well. Options protect against adverse price movements without impinging upon the possibility of getting much better prices.

Options at a glance

- The person who buys an option (put or call) *pays* the premium.

- The person who writes an option (put or call) *collects* the premium.

- A call option is an opportunity to *buy* a given commodity futures contract at a specified price and time in the future.

- A put option is the opportunity to *sell* a given commodity futures contract at a specified price and time in the future.

- The strike price is the specified price in the future.

- The contract month is the specified time in the future.

- A person exercising a call option is deciding to be long at the strike price in the *same contract month in futures* as the month for which he bought the call.

- A person exercising a put option is deciding to be short at the strike price in the *same contract month in futures* as the month for which he bought the put.

- If exercising an option would cause the holder to lose more than the premium that was paid, then he can walk away from the option, or abandon it.

Selling, or writing, an option

- The person who sells, or writes, an option, is granting an opportunity to the buyer of the option. This could be an opportunity to buy product at a specified time and price with a call, or to sell product at a specified time and price with a put. The option seller is selling these opportunities to the buyers in return for the premiums collected.

- The buyer pays the premium. The seller, or writer, collects the premium.

- The most the buyer of an option can lose is its premium.

- The most a seller can lose is unlimited.

- The most a buyer can make in profits is unlimited.

- The best a seller can do is to make or earn the premium.

The first question that springs to mind is, why would anyone accept limited gain for unlimited potential loss? They are willing to undertake this because the limited gain is more consistent over time. The writers or sellers of options make money more often than the buyers in options. Their gains are small and regular, while their losses can sometimes be large.

The reason it is called *writing* an option is because it is so similar to writing an insurance policy. The writer collects the premiums and pays out of those the casualties. With enough capital, the writer of both instruments makes money over time. Options writers, like insurance companies, try to have a pool of payers to offset losses. Options writers balance their risks by writing both puts and calls at many different numbers.

The second question one may ask is, how are premiums determined? In options on futures, it is not done with actuarial tables but with *black boxes*. These are used to calculate a premium based on the option's strike price in relation to the market, or the instrument's intrinsic value, if there is any, its time value, and its volatility factor. (Recall that the time value is the amount of time left before the option becomes due, and the volatility factor is the amount of movement seen recently in the underlying futures contract.) These three inputs, intrinsic value, time value, and the volatility factor, are run through a series of algorithms in the program of the black box, and a fair value is computed. That gives buyers and sellers an idea of the theoretical price at which the instrument should be priced.

How prices are made in options

Market-makers use these fair values to form a bid and offer, or a buying price and a selling price (for either the put or the call). If the fair value for a two-month call option priced at-the-money is 2.00¢/gal, the bid may be at 1.85¢/gal and the offer may be at 2.10¢/gal. If a seller wants to sell this instrument right away, the price it is sold at is determined by the best available bid. If it is 1.85¢/gal, that means that the best price it can be sold at is 1.85¢/gal. Someone who wants to buy it is looking for the best seller out there to be matched up with. In this case the best offer is at 2.10¢/gal. That is the price that must be paid to buy the option. They could go through the same process that Bob and Susan did above, but the market may not be as liquid. The bid and offer above may be the best out there.

When one writes an option, one is the seller of the option. That does not mean that one wants to sell the underlying commodity. Someone who writes a put is selling the option to someone else, who then has the right to sell at a specified time and price.

Buying or selling at the current bid or offer would be considered buying or selling at-the-market. If these prices are not acceptable, a *limit* or *resting order* can be placed that would become the best offer to sell or the best bid to buy. For instance, perhaps a person is looking to buy. This person could tell his broker to place a bid at 1.90¢/gal, which would turn into the highest bid or best buyer out there. If someone is in a hurry to sell, they may *hit* this bid and sell at the price that the first person liked. Of course, a person may never get filled or see the order executed. Other traders just may not be willing to sell lower than 2.10¢/gal. That may remain the best offer for a while but then get *lifted* by an aggressive or eager buyer.

The next best selling offer may be 2.20¢/gal. Another person who is eager to buy may feel that price is good and may lift the offer at 2.20¢/gal. That then becomes the market price, or last price traded. Markets move higher when buyers are eager enough to reach up and lift offers, and markets drop when the sellers are motivated enough to sell for less and hit the bid under the last traded price.

Example

The Lincoln-Lee Railroad Company needs to protect itself against the possibility of diesel prices moving higher over the next six months, and it decides to hedge itself against such a contingency. It decides to lock in a set differential against the NYMEX for the diesel it will need at its two storage terminals in Gettysburg, PA, and Chancellorsville, VA. Its suppliers agree to supply it at its terminals at 3¢/gal over the NYMEX price. That takes care of its basis risk. To hedge itself against underlying price risk, it buys call options for the anticipated needs it will have between July and December.

Table 5–2 shows the futures price and the premium price for a call option on that futures contract for each month. For sake of illustration, the strike price has been chosen closest to the settling level on the starting date of April 1.

Table 5–2 Lincoln-Lee Railroad Co. call options

Month	Futures[1]	Strike[2]	Premium[3]	Basis[4]	Effective[5]
July	53.00	53.00	2.50	3.00	58.50
August	53.60	54.00	2.50	3.00	59.50
September	54.40	54.00	3.00	3.00	60.00
October	55.20	55.00	3.50	3.00	61.50
November	56.00	56.00	4.00	3.00	63.00
December	56.60	57.00	4.50	3.00	64.50

[1] The contract price on April 1.

[2] The strike price of the call option purchased.

[3] The cost of buying the strike price call for the month listed.

[4] The differential between the NYMEX and the two terminal locations (Gettysburg & Chancellorsville). In this case, Lincoln-Lee has preagreed with its supplier to fix that differential at 3¢.

[5] The total price that Lincoln-Lee will pay for its material at its two terminals.

All amounts listed in ¢/gal.

In this case, it makes no sense to exercise an option to buy July futures at 53.00¢/gal when the prevailing market value is only 50.00¢/gal. Lincoln-Lee lets the July 53.00¢/gal call option expire without exercising it, and the company saves itself 3.00¢/gal.

Table 5–3 How Lincoln-Lee used its July call option

Month	Futures[1]	Premium[2]	Basis[3]	Effective[4]
July	50.00	2.50	3.00	55.50

[1] The contract price on April 1.

[2] The cost of buying the strike price call for the month listed.

[3] The differential between the NYMEX and the two terminal locations (Gettysburg & Chancellorsville). In this case, Lincoln-Lee has prearranged with its supplier to fix that differential at 3¢.

[4] The total price that Lincoln-Lee will pay for its material at its two terminals.

All amounts listed in ¢/gal.

In late June. As time passes, Lincoln-Lee will have choices to make. These can be addressed individually. By the end of June, July heating oil futures have fallen 3.00¢/gal. Table 5–3 shows how this information will now look. The strike price has been removed because it is no longer desirable to buy at that price or to exercise the 53¢/gal call option purchased above. The premium remains in this table, because the premium has been paid already, and it is now a part of the overall cost. The basis is still secured with Lincoln-Lee's supplier. Consequently, if the company goes to buy a futures contract at the existing market price in late June, it will pay 50.00¢/gal plus the locked in basis of 3.00¢/gal. The cost of the premium it has already paid must also be added in.

Had Lincoln-Lee not purchased the 53.00¢/gal call on April 1, its effective cost would not include the 2.50¢/gal premium it paid. However, the company would be at the risk of potentially having to pay more had prices advanced. This is what they in fact do through the month of July used in this example. The calls it purchased protect the company against higher prices without taking away any potential benefits of lower prices.

In late July. By late July, the August contract has rallied back most of the distance it had fallen in June. When the decision is made, August is trading at 53.50/gal, 10 points lower than where it had been on April 1 (see Table 5–4). (It was at 53.60/gal on April 1, in Table 5–2.)

Table 5–4 How Lincoln-Lee used its August call option

Month	Futures[1]	(Strike)[2]	Premium[3]	Basis[4]	Effective[5]
August	53.50	(54.00)	2.50	3.00	59.00

[1] The contract price on April 1.
[2] The strike price of the call option purchased.
[3] The cost of buying the strike price call for the month listed
[4] The differential between the NYMEX and the two terminal locations (Gettysburg & Chancellorsville). In this case, Lincoln-Lee has prearranged with its supplier to fix that differential at 3¢.
[5] The total price that Lincoln-Lee will pay for its material at its two terminals.
All amounts listed in ¢/gal.

The strike price was 54.00¢/gal, which has been left in the table in parentheses to signify that it is no longer the operative buying price. Lincoln-Lee can buy futures at 53.50¢/gal. It then can add in the cost of the premium it had already paid and the basis differential it had agreed upon with its supplier to reach an effective price of 59.00¢/gal. If it exercises the call option, it will be long at 54.00¢/gal instead of at the prevailing market price of 53.50/gal, losing 0.5¢/gal.

In this case, as well, Lincoln-Lee decides *not* to exercise its call option.

Table 5–5 How Lincoln-Lee used its September call option

Month	(Futures)[1]	Strike[2]	Premium[3]	Basis[4]	Effective[5]
September	(55.50)	54.00	3.00	3.00	60.00

[1] The contract price on April 1.

[2] The strike price of the call option purchased.

[3] The cost of buying the strike price call for the month listed.

[4] The differential between the NYMEX and the two terminal locations (Gettysburg & Chancellorsville). In this case, Lincoln-Lee has prearranged with its supplier to fix that differential at 3¢.

[5] The total price that Lincoln-Lee will pay for its material at its two terminals.

All amounts listed in ¢/gal.

In late August. Even though Lincoln-Lee did not recoup the full value of the premium it paid (3.00¢/gal), it still did manage to pay less in this case by exercising its call option than by letting it expire. However, for each of the three months so far reviewed, Lincoln-Lee would have been better off, as things worked out, by not hedging through call options. In each case, the premium, either in full or in part, raised the effective price that Lincoln-Lee paid.

Because the premium had already been paid, what Lincoln-Lee did here was to defray part of that expense by exercising its call option.

By the end of August, prices have continued to advance, with the September contract now trading at 55.50¢/gal. If Lincoln-Lee buys a September futures contract on the market, it will have to pay 55.50¢/gal. When one adds in the cost of the premium and the basis, the effective cost would come to 61.50¢/gal. Instead it decides to exercise its call at 54.00¢/gal, saving itself 1.50¢/gal (see Table 5–5).

Hedging does add costs to doing business, and in a perfect world where prices remained constant, there would be no need for it. Hedging should be viewed as an insurance policy. Sometimes it is better to be out the premium than to collect on the policy. But without the policy, the risk is not addressed.

In late September. As time advances, so do prices, and by late September, October futures have risen to 65¢/gal. At this point, Lincoln-Lee is glad that it hedged by buying call options. It is holding a 55¢/gal call, which it will happily exercise.

Table 5–6 illustrates what Lincoln-Lee decided to do and how this affected its bottom-line.

Table 5–6 How Lincoln-Lee used its October call option

Month	(Futures)[1]	Strike[2]	Premium[3]	Basis[4]	Effective[5]
October	(65.00)	55.00	3.50	3.00	61.50

[1]The contract price on April 1.
[2]The strike price of the call option purchased.
[3]The cost of buying the strike price call for the month listed.
[4]The differential between the NYMEX and the two terminal locations (Gettysburg & Chancellorsville). In this case, Lincoln-Lee has prearranged with its supplier to fix that differential at 3¢.
[5]The total price that Lincoln-Lee will pay for its material at its two terminals.
All amounts listed in ¢/gal.

Lincoln-Lee exercises its call option to buy futures at 55.00¢/gal. It keeps an effective price of 61.50¢/gal (55.00¢/gal for futures, 3.50¢/gal in premium, and 3.00¢/gal in basis). Had it not hedged, it would not have laid out the premium, but it would have paid a market price of 65.00¢/gal plus the basis of 3.00¢/gal to yield an effective cost of 68.00¢/gal. By hedging, it saved 6.50¢/gal for this month.

In late November. Prices ended up peaking in October at slightly higher levels than they had reached in late September (see Table 5–7). For sake of argument, November futures have the same profile in late October as October futures had in late September. The company once again saved money by exercising its option. Prices drop sharply in November, though, and by the end of that month, December futures are trading at 46.00¢/gal. Warm temperatures, high refinery runs, and international tranquility push prices lower. With the market now at 46.00¢/gal, Lincoln-Lee abandons its option and pays the market price.

Table 5–7 How Lincoln-Lee used its December call option

Month	Futures[1]	(Strike)[2]	Premium[3]	Basis[4]	Effective[5]
December	46.00	(57.00)	4.50	3.00	64.50

[1]The contract price on April 1.
[2]The strike price of the call option purchased.
[3]The cost of buying the strike price call for the month listed.
[4]The differential between the NYMEX and the two terminal locations (Gettysburg & Chancellorsville). In this case, Lincoln-Lee has prearranged with its supplier to fix that differential at 3¢.
[5]The total price that Lincoln-Lee will pay for its material at its two terminals.
All amounts listed in ¢/gal.

In this case, Lincoln-Lee walks away from its option and buys futures at the prevailing market level of 46.00¢/gal. Adding in the cost of the premium and the basis, Lincoln-Lee pays an effective price of 53.50¢/gal, saving itself 11.00¢/gal from its projected costs.

One could make an argument that it would have been cheaper to forego the entire process. And there may be validity to the theory that it will all even out over the years. Of course, the question is, is one ready to take that risk this coming year? In reality, some sort of hedging is in any business's best interests. Options or futures may not be the answer for everyone, but they are viable instruments that can be blended together with basis protection to provide good protection against adverse price moves. Options may cost more, but they do provide protection in both directions. If there is a need to buy in the future, a call option will limit how high a price one will have to pay, but it will not limit the downside. One can walk away from a call option in a declining market and pay the lower market price.

Example

The Wylie Roadhauler Corp. has a fleet of trucks. It has contracted with ACME Corp. to transport goods from Cleveland to Tucson for the next six months. Other logistics, such as the return run, or what the Wylie trucks may be carrying back from Tucson, will not be considered in this example, unrealistic though that may be. It is a long haul, and Wylie is going to have to be able to line up suppliers at some of the major hubs that its trucks will pass through, like Cleveland, St. Louis, and Tulsa.

For further purposes of illustration, it is assumed that Wylie has worked out a steady differential against NYMEX prices with refineries at those three major hubs, which will be able to cover the distance. Clearly this has been simplified a great deal for sake of clarity, and the reality is obviously much more challenging. But Wylie Roadhauler Corp. needs to give ACME a price on how much it is going to cost to do the business. These are the inputs.

Wylie Roadhauler Corp.'s operating framework. In order to give ACME a price, Wylie must figure out its costs, which are determined by the following equation:

Total Costs = (gallons used x diesel cost) + driver cost + profit margin

The cost of the driver and the profit margin are fixed variables. Diesel cost is determined by adding the following:

- Gallons used from Cleveland to St. Louis x (Cleveland differential + refiner fee + NYMEX price)

- Gallons used from St. Louis to Tulsa x (St. Louis differential + refiner fee + NYMEX price)

- Gallons used from Tulsa to Tucson x (Tulsa differential + refiner fee + NYMEX price)

- All relevant taxes and fees

The differentials between the NYMEX and the various hubs have been fixed in advance, as suggested above. In addition, the mileage and the fees that each company will charge for diesel supplies are known. Thus, the only variable left to lock in here is the underlying price, as established by the NYMEX futures. Some trucking companies are better off getting the price from their suppliers, who will then lock it in on the futures. But for the sake of understanding what is going on, the calculations will proceed as if Wylie were a seasoned and well-informed user of futures and options.

Wylie's management and its head trader feel that prices may drop first, but they have to get out a hard dollar figure to ACME. They decide to use call options. They include the premium price in their cost of doing business, and if prices rise, they are protected. If prices drop, they can augment their bottom-line, or they can refund a portion or all of the savings to ACME to insure further goodwill. This is a judgment call, a business decision to be made if and when any savings are realized. Either way, the business relationship will make sense for both companies.

Wylie pays 3.00¢/gal to purchase a 70.00¢/gal call. That makes its effective cost 73.00¢/gal, plus the set refiner fees, times the number of gallons it will use in the run from Cleveland to Tucson.

Example

There is at least one other way that options can be used. Perhaps Grandfather's Home Heating Corp. wants to hedge some October heating oil, but is not as eager for it as it is for December or January gallons, because October is not generally as cold. A fewer number of gallons will be needed in October, and Grandfather's is not likely to be as much on the line. Rather than buying a call to protect against higher prices, Grandfather's decides to write an October at-the-money put and collect the 2.00¢/gal premium.

When one writes or sells a put, one is in effect looking for higher prices. If prices do move higher, the put will not be exercised, and the writer or seller of the put can keep all of the premium.

What they are doing is collecting a premium and saying that if the price drops, they will allow the buyer of the put to be short at the strike price. (If the market was trading at 52.00¢/gal when the put was sold, then that would be the at-the-money strike price at the time.) Once prices fall beneath that level, Grandfather's will be allowing the purchaser of the put option the right to be short at 52¢/gal (sell it at 52¢/gal). Who will the purchaser of the put sell its futures contract at 52¢/gal to? The buyer is Grandfather's Home Heating Corp. Now Grandfather's is long, by selling a 52.00¢/gal put, and having the long October futures contract put to them at 52.00¢/gal.

When one writes or sells a put, one is in effect looking for higher prices. If prices do move higher, the put will not be exercised, and the writer or seller of the put can keep all of the premium.

However, that is not the actualized price that Grandfather's pays. Because Grandfather's collected a 2.00¢/gal premium, that must be subtracted from the final purchase price to give an actual purchase price. In this instance, it would be 52.00¢/gal (at-the-money put strike) minus 2.00¢/gal, or an even 50¢/gal. Since Grandfather's wanted some product for October, this was just one more way of getting it, while saving 2¢/gal on the deal. If the purchaser of the put had let it expire without exercising it, Grandfather's would have still made 2¢/gal, although it may have had to pay a higher price for its product.

This is not a method to use in cases where risks are high. There was never any guarantee that prices would fall to the point that product would be put to Grandfather's. It may have just made 2¢/gal and then watched prices shoot up 25¢/gal. This was just another illustration of how things could work.

USING AN EXCHANGE FOR PHYSICAL (EFP)

There is one other method of oil trading that is commonly used. It involves the use of futures and basis protection, and it gives greater flexibility to the long hedger. As a drawback, it also can incur margin calls, or a steady outlay of cash. This is not recommended for traders who are either unfamiliar with futures and how they work or for companies that do not have ready access to potentially sudden demands for cash.

Here is how it works. First, a company will lock in its basis under the rack with a supplier. In this case, it has to be a supplier who uses futures traded on the NYMEX. The difference between what futures are and what the energy user will pay under the rack is agreed upon this way in advance.

It is now up to the buyer and seller to fix their own underlying prices (for crude oil, heating oil, gasoline, etc.) through futures, as and when they see fit. They can do that at different and separate times and prices. At some point immediately before the actual transfer of physical material is made, the two parties will cancel out each other's futures transactions with an EFP. The would-be buyer has a long futures contract and the supplier has a short futures contract. These two positions are brought together through the NYMEX, and they offset each other. They are replaced by cash or physical commitments, with the long futures holder exchanging his long futures position for a long cash position. The supplier exchanges his short futures position for a short cash or physical position—a commitment to deliver product.

The two parties to this transaction may agree in June to undertake this EFP in December. Between June and late November, they will establish their corresponding futures positions. The buyer will buy futures, and the supplier will sell futures. These parties do not need to initiate their positions at the same time. The energy user may buy December heating oil futures in late June, for instance, at 55¢/gal. The seller, in this case the supplier, may wait until October to sell December heating oil futures, assumed here to be at 65¢/gal. On the day before Thanksgiving (or thereabouts), the two parties may agree to do an EFP. For purposes of this example, it is assumed that the price at the time is 60¢/gal, and both parties are agreeable to using that price.

In actuality, any price can be used, which is a point that may require explanation. If the supplier wants to use 70¢/gal or even 80¢/gal as the EFP price, even though prices are trading at 60¢/gal, and the buyer agrees, the end result will come out the same. The NYMEX allows the two parties to agree upon any price they wish. It does not matter. If the price is high, the supplier will take a futures loss that will be offset by a physical gain. If the price is low, the buyer may take a futures loss but will pay that same price plus the prearranged basis differential for the actual wet barrel.

The example calculations can be continued using 60¢/gal for the price.

The buyer

- The buyer bought a December futures at 55¢/gal (to lock in a buying price).

- The buyer sells that December futures contract (paper instrument) through the EFP at 60¢/gal, making 5¢/gal profit on the futures.

- Then the buyer pays the agreed-upon 60¢/gal plus the prearranged differential for the physical product. After factoring in the 5¢/gal futures profit (on the paper instrument), the buyer pays 55¢/gal plus the differential already agreed upon.

The seller

- The supplier sold futures at 65¢/gal (to lock in a selling price).

- The seller buys the EFP at 60¢/gal, making 5¢/gal profit on the futures position. That is the same as buying a futures contract and switching it with a short cash position.

- The seller then receives 60¢/gal plus the prearranged differential for the physical product. After factoring in the 5¢/gal futures profit, the seller gets 65¢/gal plus the differential already agreed upon.

An EFP allows traders to exchange paper futures positions for wet, physical positions. The holder of a long futures position offsets it at the agreed-upon price and gets physical product from the seller. The holder of a short futures position offsets it at the agreed-upon price and delivers physical product. The EFP allows two physical traders who do business with each other in the real world to offset their paper futures positions and segue straight into the physical deal.

EFPs allow traders to have the financial security that comes with exchange-traded contracts while still allowing them to trade cash with each other. It also lets buyers and sellers set their prices independently, rather than at the same time. They know that they are going to make or take delivery from each other, but they have different ideas about when it may be best for them to set their prices.

As can be seen, this method allows both players to time their initial futures transactions to get the best prices for their companies. The buyer gets an underlying price of 55¢/gal in June, while the seller gets a higher selling price of 65¢/gal in October. By using the EFP, both parties get the best of both worlds and are left whole and hedged after all the transactions are completed.

Had both parties agreed upon an EFP price of 75¢/gal, it would have worked out exactly the same. For example:

- The buyer bought a December futures at 55¢/gal.

- He sells it through the EFP at 75¢/gal, making 20¢/gal profit on the futures.

- He then pays 75¢/gal plus the prearranged differential under the rack. The next step is to factor in the 20¢/gal futures profit, and the buyer pays 55¢/gal plus the differential already agreed upon.

- The supplier sold futures at 65¢/gal.

- He buys it through the EFP at 75¢/gal, yielding 10¢/gal loss on the futures.

- He then receives 75¢/gal plus the prearranged differential under the rack. The next step is to factor in the 10¢/gal futures loss, and the seller gets 65¢/gal plus the differential already agreed upon.

No matter what price is used for the EFP, both buyer and seller get their corresponding underlying price (established by futures), plus or minus the prearranged differential.

When to Use Futures and Options

The first consideration for any real world hedger is the basis risk between cash or physical barrels and futures. If there is a history of seasonal volatility, futures or options can hedge only a part of the risk.

Futures and options are fine to use during periods when there is little risk of supplies not being available. This normally would be the case for heating oil distributors, diesel resellers or marketers, airlines, railroads, or other end-users during the summer months in locations hooked up to New York Harbor. It would also be the case for gasoline marketers or end-users during the winter, again if they use terminals in or around New York Harbor. Many locations in the Northeast and Mid-Atlantic states would tend to fall in this category, while the Midwest does not. California definitely does not.

One needs to study the parameters of trading between local racks and futures prices. This information can be obtained from various news and price reporting services and will help to assess risks in a given location. Knowing when cash prices are likely to be high or low against futures, based on historical figures for a given region, is critical.

Here are some useful rules of thumb:

- **Buyers.** Buyers *should not* buy futures when local racks (cash prices) are at historically high discounts to New York. Buyers *should* buy futures when local racks are at historically high premiums to New York. When racks are at high levels relative to futures, it is possible for cash to fall and futures to rise, over time. When cash is at historically low levels relative to futures, it is more likely that cash prices will rise more quickly than futures. The two will try to converge.

- **Sellers.** The inverse is true for those holding inventories. Sellers *should not* sell futures against inventories when prices are at historically high premiums to New York. Sellers *should* buy futures against inventories when prices are at historic discounts.

CHOOSING BETWEEN FUTURES, OPTIONS, CAPS, AND FIXED-PRICE PROGRAMS

Deciding between futures and options or between fixed wet barrel programs and capped (wet barrel) programs is a matter of how fierce the competition is in a specific locality. If it is extremely competitive in the spring or summer, when customers are shopping for a winter fuel price, then paying the premium to get an option or a capped price program may be too steep to afford.

Of course, that can be a two-edged sword. Prices might fall, and the competition could either not be hedged or be hedged less. Then purchasing fixed-price protection could result in a less advantageous position relative to the competition. Customers may decide to switch to a competitor in the middle of the season if prices decline. If customers want the lowest price when they agree to it, using fixed prices may be best. If customers want the best price when they actually burn the product, capped price protection may be better.

Example

Liberty Island buys its oil from one of two distributors, Empire Standard Oil or Garden Variety Oil. Empire likes to use capped price programs (buys call options), while Garden likes to use fixed price programs (buys futures). They both agree that prices are a buy at 50¢/gal (on the NYMEX) for December barrels. When they buy, there is no differential or basis with New York Harbor. Their suppliers charge them 3¢/gal more than the screen for their profit margin, and to lock in a fixed price costs an extra 1¢/gal. To lock in a capped December price, it is an extra 3¢/gal.

In this instance, Empire has a *maximum* price of 56¢/gal (by using call options), while Garden has a *set* price of 54¢/gal (by using futures). If prices move higher, Garden will maintain a fairly steady 2¢/gal advantage over Empire. If prices drop under 53¢/gal, then Empire will start to experience a numerical advantage over Garden. However, it would always be 3¢/gal higher than a company that did not hedge at all. Of course, both Garden and Empire would have arithmetically increasing advantages over this third company in a bullish market of any significance. Sometimes, companies would be better off without hedging at all.

The difference is really a matter of how much downside risk there is compared to the price sensitivity in the local market. If customers will switch to the competition for the cost of the premium, then the chances of a sharp downturn must be considered. Those chances will be dependent upon the market's fundamental, technical, seasonal, and psychological outlooks, which are then modified for price level. Clearly, anyone would feel safer locking in a set price of 40¢/gal than one of 95¢/gal.

Fixed-price programs or futures work well when customers are unlikely to leave if prices drop during the time in which they will actually pay for product. Capped-price programs or call options work better when there is little customer loyalty, and customers will switch to save a few cents. Fixed-price programs and futures are good at historically lower prices, while capped-price programs and call options work better at historically higher prices. Fixed-price programs and futures give a hard-and-fast price, and they protect against the upside only. Capped-price programs and call options protect against both higher and lower prices. That is why they cost more.

In recent years, it would have been unwise to lock in fixed prices during the spring for the following winter. In both 2003 and 2004, March was not a good time to lock in prices, because they were at historically high levels. It would have been better to cap prices or wait.

The Midwest: special problems and opportunities in refined products

Diesel or gasoline resellers in the Midwest need to use futures or options in combination with basis protection to serve their purposes. It is also advisable to have a good idea of the trading parameters of a given location in relation to New York. If local prices are near the bottom of their traditional parameters against NYMEX prices, then it is *not* a good time to use these instruments. New York's prices could remain flat, while one's local prices move above New York's prices into a premium relationship. The result is higher prices under one's racks but no gain in futures. Instead it makes more sense to prepurchase supplies from a local supplier, or to place supplies in storage and sell New York against any inventory risk one may have as a result.

It makes sense to buy futures when Midwest premiums against the NYMEX are at or near parameter highs. In that manner, one can make money both ways. Racks could drop against New York, while New York advances. The possible result could be lower buying prices under the rack and a profit on the futures.

If a producer is relying on just one supplier, or on suppliers getting their supplies, in turn, from only one source (such as one refinery), it would be wise to look for a backup. When any refinery in the Midwest goes down, or a unit goes down, it can create special problems that are best avoided. *Futures and options only protect against price risk.*

Crude oil

Refiners most frequently will want to buy crude oil futures and sell refined products futures at the same time, in order to lock in a refining margin. There are times, though, when buying crude oil calls at historically low prices, or buying refined products puts at historically high prices, will make sense.

Refiners can also sell fixed-price programs to resellers to lock in additional margins. In those cases, the refiner only needs to lock in the crude oil price. When selling a capped-price program, the refiner can either buy crude oil calls or can buy the crude outright and buy refined products puts to protect against the downside. Complicated strategies can be enlisted by refiners to maximize profits.

Producers have a special range of opportunities that comes with the ability to look much farther ahead. They can buy puts when prices are high, or sell calls at a number of prices. They can also use futures and options in combination. Producers should enlist the help of a specialist to customize a useful hedging program that suits lenders and shareholders alike.

Selecting a futures broker

There are a few very expert groups of futures brokers out there who are adept at guiding hedgers through the potential pitfalls of futures trading. Most of these brokers are specialists who have been working with hedgers for many years. All of them have hedging clients now. It is important to ask for references and not let the name they are affiliated with be a big influence. It is a specialty field, and anyone who has been in it for a long time will have been forced to change companies several times just because of mergers and buyouts. These brokers should be sought out. They know much more about their clients' needs and worries than a stockbroker at the local, big-name commission house. Often they will charge less, as well.

It would not be wise to use a personal stockbroker or an existing financial relationship if the broker does not have *experience and existing accounts with similar needs.* No one wants a stockbroker discussing hot tips when there is a need to hedge oil or natural gas. By the same token, it would not be wise to use a grains and fibers expert when an oil specialist is available.

On a final note, a hedger should tell his broker that he is not immediately interested in any *opportunities*, just the implementation of a sound hedging plan. A person who wants to speculate should keep separate accounts and not get the two confused. In a well-conceived hedging program, one should not need to buy and sell more than a couple of times a month, or perhaps each week in winter.

Cameron Hanover clients are advised to buy once for the winter. If a top comes along in the meantime, clients might be encouraged to buy put options or sell futures against the hedge. That is as fancy as the program gets. Usually clients are encouraged to buy the puts so that the upside protection is not negated, as it can be with futures. There are, on average, four major trends per year in energy markets. A hedger who does too much fine-tuning can end up fine-tuning himself out of a good hedge. If someone catches two trends and avoids being hurt by the other two, that person will have had a good year.

This can be summed up easily. Offense sells tickets, but defense wins championships. The producer should never call his broker for excitement, only for sound hedging advice or implementation. A business that operates profitably will allow its owner the purchase of season tickets to a favorite local team's game or a vacation in Atlantic City.

A person may also want to get advice from one source and implement it through another. There are several hedge advisory services that do not make money through transactions. They get paid monthly subscription fees and let people choose their own brokers to implement the trades. By staying away from transaction fees, these advisory services do not have any interest in persuading a person to trade.

Bank Swaps and Inventory Financing

Banks and suppliers can offer private financial instruments that can act like either futures or options, or like physical contracts. Some brokerage houses offer them as well. The list of institutions that offer them is constantly changing, and it is important to understand exactly how these instruments can or cannot protect one from specific risks. One needs to address each of the risks previously outlined.

Some of these programs work beautifully, and some offer inventory financing and guaranteed supply. However, these programs rely entirely upon an understanding and a level of sophistication that demand dedicated talent and a commitment to continuity. Many banks have not made the proper commitment to this area or are uncertain how to respond to margin calls resulting from their own exposures being hedged. Metallgesellschaft, the huge German bank, ran the most sophisticated and innovative hedging programs in the world, until it had to finance its own exposures. Then it panicked and fired a staff of the most talented and experienced heating oil people in the United States. The continuity was gone, and so was the expertise. They were replaced with people talented in another field—the liquidation of companies and exposures.

The programs died on the vine because German bankers could not grasp the complexities of hedging heating oil in the United States. Maybe they should not be expected to. But they had hired the best in the business and then panicked at the very worst time. Had they trusted the talent they had assembled, all of the positions would have returned profits, resulting in millions in profits for the bank. Instead, they amassed huge losses and provided the press and the locals some amazing opportunities.

Like a dueler with a dead eye, Metallgesellschaft saw only one side of the balance sheet and then lost nerve. What they could not see was the incredible penetration their subsidiary had made into the U.S. market. They had been scared by an afternoon shadow of their own success!

In the process, certain hedging procedures have been tarred with the same brush. It is important to know the people and the organization with which one is working. In contrast to the example above, some companies have been in business long enough to trust, and they can construct very specific hedging programs that are safe and sound. Anything can change, but this is a book about tomorrow based on what is known and available today.

This is a natural business for banks to be in. Unfortunately, it is not a natural one for bankers to be in. Most bankers have a mastery of certain mainstream banking concepts, but attaining this knowledge has left them with insufficient time to gain the specific expertise needed in this field. They know a lot banking but not enough about *hedging energy*.

The most important factors involved with these instruments are expertise and continuity. Some companies have made it a point to understand what they are doing, and they do it well. One is looking for talented people who understand oil and have the backing of the financial institution. Sometimes the institution has done this kind of work before. Sometimes, it is new for the financial group, but not to the people they have hired. One should focus on the people they have hired.

Timing

Pure hedging, or the theory of hedging, was addressed in an earlier chapter, and for those without any experience or guidance, it is best to start without any bias on the market's direction. But as one becomes more familiar with the market, the desire to try to fine-tune one's hedging operations will become overwhelming. The business world is just too competitive for one not to try. Every hedger will eventually wind up trying to balance on the razor's edge between hedging and speculating. This dangerous phase must be faced honestly so that a trader does not overreact, undoing years of patient learning.

Beware of overtrading.

One way to deal with this problem is to have two different accounts. This way a person can keep track of the purpose of each trade. One of the important aspects of intelligent hedging is the ability to compartmentalize. The successful hedger needs to be able to differentiate between conflicting objectives. A heating oil distributor may be long February futures or call options or wet barrels to protect against higher prices. But, in the final days of December, he may also be short futures against LIFO (last in, first out) barrels that have been purchased for tax purposes. (These are wet barrels that he has put in storage.) Different objectives can bring opposing responses. Many hedgers go wrong when they lose track of their objectives.

Beware of confusing hedging with speculating.

That does not mean that one cannot use one's knowledge of current conditions to time an entry or exit from a hedge. In today's market, with cut-rate competition, a good knowledge of current conditions is vital to remaining competitive. Even when prices are extremely low, it is important to take into account what the competitors will offer if prices drop even more.

DECIDING WHY AND WHEN

The hardest part for someone coming into this for the first time is getting past the *why*. Traders coming into this market without any prior background in trading are at a loss to figure out how all these other people are making split-second decisions without anything new or major coming into play. There is no fresh news, and yet the activity in the ring is vigorous and hectic. Why are these people buying or selling? That is a question frequently asked by people starting in this business. And it is a difficult question to answer. It is very much like the blind men describing an elephant when each one is holding a different part. The trunk is quite a bit different from the ears, legs, back, or tail.

The closest answer is that it is somewhat similar to the process that goes on at a cafeteria for lunch. Some people are eating things they like, some people are eating things that will help them lose weight, and some people are eating things that do not cost as much. There are innumerable other reasons that someone might or might not be eating something.

There are similarities in trading. Some people, like the independent floor traders, or *locals* as they are called, are looking to get in and out of the market in the same day. Some active speculators will be looking to get in for a few days, and some funds will be thinking along the lines of one week or even one to two months. Commercial traders may be putting on positions all the way into the very distant future. One can think of these as the quick eaters or the slow eaters. What they order in the cafeteria may be dependent upon the amount of time they have or do not have available. In trading, some people think in terms of small profits generated often, while others think in terms of bigger profits generated less frequently.

People also have different risk parameters. The quick trader cannot let losses become excessive. The long-term trader may have a bigger acceptable risk because of higher profit expectations. *The general rule of thumb is to have potential expected gains equal to three times the acceptable risk.* While this rule applies primarily to speculators, it also is useful for hedgers to know.

One liquidates an existing long contract or covers an existing short contract.

In roughly half of the volume for any given day, half the trades will be made to take a loss before it gets worse. That is a simplification, but *for every dollar made, there is a dollar lost in commodities.* This is not the stock market where a rising tide lifts all ships. This is the kind of business where it is not enough to succeed, but where others must fail. That may be putting too sharp an edge on it, but the basic premise is there.

Markets get moving fastest when people are getting out of losing positions. If one does not get filled on a fresh position, one missed an opportunity. If one lets a bad position get worse, one will be lining up margin money tomorrow morning. This feature has a certain clarifying value, like smelling salts, and it tends to make people losing money a little bit more involved in the price quotation process. Somebody losing money in an open position is more likely to call his broker every few seconds. The sweat drips down his forehead as he clings to the phone like a lifeline. The person who wants to get into a new position but misses it can shrug it off. The person in a losing position that is getting worse can smell his own blood and see circling dorsal fins.

> **"He who sells what isn't his'n**
>
> **Pays the price or goes to prison."**
>
> **—an old trading adage**

Usually, people holding losing positions start off trying to minimize the loss by placing a *limit order*, which is a specific price above the market when one is selling or below the market when one is buying. This is placed in lieu of a *market order*, which is the best price available immediately. All this does is tip off the locals (independent speculators trading on their own hook on the floor) and lets them know where the orders are thickest.

If there is better selling over (than buying under) the current price, then the locals are happier selling, and vice versa. Eventually, often after several unsuccessful limit orders go unfilled (the price is not executed on the floor for one's account), the trader losing money will go the market. If this process happens with several traders simultaneously, as it often does, then prices can move rapidly.

There are many different reasons why traders buy and sell contracts, but the most motivated buyer is a short losing money, while the most motivated seller is a long losing money.

DIFFERENT TYPES OF MARKETS

There are four different types of markets that occur in energy markets. Each has its own distinct characteristics, and there are times when one or more will exert an influence over price movement. The following is a brief description of each type.

Fundamental

Prices can be moving in a specific direction because of a dramatic change in supply or demand. Oil prices usually move fundamentally on Wednesday morning as traders respond to the weekly American Petroleum Institute (API) and Department of Energy (DOE) inventory reports. Unscheduled downtimes at refineries, pipeline problems, the approach of bitterly cold temperatures, or changes in OPEC production will also affect oil prices.

With natural gas, prices respond Thursday mornings to the release of the weekly EIA report, which shows underground storage figures. Natural gas prices also respond to Monday morning weather forecasts and the sightings of tropical waves, storms, or hurricanes over the summer. These are capable of altering both supply and demand.

Fundamental markets are characterized by enduring trends that can take anywhere from 6 to 22 weeks to develop.

Technical

Prices can be moving in a specific direction because of price action or because of what the charts look like. The locals are almost 90% inspired by charts, and a great many brokers use them, too. So do trading funds. Technicals are explained in more detail in chapter 10.

The driving force behind a technical move is the trend. Prices went higher last week, so why shouldn't they go higher this week? That is the underlying reasoning.

Seasonal

Prices also can be moving in a specific direction because of an existing seasonal pattern. Oil prices usually reach lows in March and July and hit highs in October and November. This is exactly the opposite of what would be expected logically. It is important to be aware of these seasonal tendencies and not let logic run riot.

Psychological

Finally, prices can be moving in a specific direction because something has a psychological hold on the market. This could be a war, a strike, a coup, or speculation concerning other events. When an event affects the energy market, and its ramifications on supply or demand are way out of kilter with the market's response, then it can be described as a psychological market. It easily can be compared to a tornado.

When one is in the most dangerous of these psychological markets, the F4s and F5s (like the Gulf War), there is one certain way to know. The Exchange raises margin deposits. If this is happening regularly, then one is in a psychological market. They are virtually impossible to trade in unless one is down there on the floor or is specially gifted. Prices whip around like an untended fire hose, and they peak or bottom without warning. Prices can do in a day what once took a week or even a month.

Here is a rule of thumb: Markets always go farther and faster than anyone dares to imagine. It is important to stick with the trend. If prices seem too high or too low, they are not. One should abandon logic or reason, because the market has.

Fundamental Analysis

Fundamental analysis is the study of supply and demand. There are basically two schools of thought, or two different approaches, to analyzing the markets. The technical approach studies the behavior of prices without any regard to fundamental factors. The fundamental approach looks at relative levels of production, imports, consumption, or inventories, and by comparing these factors, arrives at an opinion about current prices.

Consider a case in which inventories are the lowest they have been in a number of years (as they were in most of 1996 and again in 2003). If demand is growing, then a fundamentalist might compare existing prices to prices of previous years. When existing prices are low, then he might conclude that prices are a good buy. But if prices are high by historical standards, the fundamental trader might decide to sell the market. At least that is how it works in theory.

In reality, more often than not, the price levels that traders are comparing current prices against are yesterday's prices, or even prices five minutes ago. Perhaps prices are trading at 60.00¢/gal for the expiring month heating oil futures on the NYMEX, and a piece of bullish news comes out. Very few traders are going to have the luxury of doing more than a cursory investigation or comparison of fundamentals. And the most common reaction will be, "It is more bullish than it was five minutes ago, so it's a buy."

These seat-of-the-pants responses will work to affect the market in combination with margin calls, technical moves, and the combined longer and shorter term responses of speculators and hedgers. Together they will push prices to levels where they may have value or be overpriced. The job of a pure fundamentalist is to find those situations where the market represents value, or a good buy, or where it is overpriced and represents a good selling level.

Generally speaking, supply and demand will exert influence on the market. However, it can take a very long time for this influence to become evident. And the concept of value in a futures market can be extremely difficult to gauge. This is also frequently confused by the supply and demand for futures contracts.

A short *squeeze* can occur when prices move higher and traders are buying futures, but there is no corresponding tightness in cash barrels. This can only last for a brief period, but these squeezes are frequently seen two to three days before a contract expires. This is at the point where traders may be worried but are still uncertain about the approaching availability of product. Usually, there is no problem, but prices can run up sharply during the time it takes for traders to find that out with any certainty.

It is important to be aware of the market's fundamentals as they change, but it is equally important not to get too far ahead. A discussion of the important fundamentals will be given, but some knowledge of technical analysis or seasonal or open interest analysis is recommended to help with timing any fundamental trade.

WHICH FUNDAMENTALS ARE IMPORTANT?

Choosing which fundamentals are important can be a highly subjective and very specialized task. However, some of the most frequently watched fundamentals will be explained, along with how they can be looked at to give a more solid footing upon which to base a decision.

The fundamental foundation of the oil complex is a combination of OPEC production and exports (which once were almost synonymous). This information is released monthly by the International Energy Agency (IEA), the DOE, and various news wires in the first two weeks of each month (for the preceding month). It is also released in the weekly API and DOE reports. Knowing these numbers and understanding how they relate to the oil complex will allow a useful background picture of the overall market.

OPEC and non-OPEC

OPEC numbers are important because the *incremental* barrel comes increasingly (almost exclusively) from OPEC's 11 member countries. There are 5 countries (the U.A.E., Kuwait, Iran, Iraq, and Saudi Arabia) that currently hold almost all of the incremental barrels and have 65% of the world's proven oil reserves. The other members of OPEC are Venezuela, Libya, Algeria, Nigeria, Indonesia, and Qatar.

Proven reserves are those barrels that can be recovered at existing prices. As prices increase, so do proven reserves.

Non-OPEC producers such as Russia, Britain, Norway, Mexico, Oman, Egypt, Malaysia, and Yemen comprise a second tier of swing or incremental producers. However, none of these countries, either singly or in combination, has the proven reserves or infrastructure in place to increase production as dramatically as OPEC has as a group. And within OPEC, the countries from the so-called Petroleum Gulf and Venezuela have the reserves to do it on a sustained basis. The other OPEC members do not have the oil in the ground or as strong a likelihood of recovering those reserves. As prices increase, the ability to get oil increases.

To subdivide the Big 6 even further, there are the Saudis and the Venezuelans, who have the barrels and the technology and investment to economically recover them. The U.A.E. and Kuwait have substantial access to capital and technology but lack the motivation to increase production as much as the first two countries.

There are also the countries of Iran and Iraq. These countries have larger populations and faster-growing domestic consumption than the other four countries. Both have had limited access to Western capital and technology. They both would like to increase production but have been strapped by cash needs elsewhere. Iraq is emerging from years of neglect and abuse at the hands of Saddam Hussein. It remains to be seen how investment and democracy will shape the future of the country's oil capacity.

Russia has become the fastest growing producer outside of OPEC, and investment in its oil capacity has paid huge dividends. As of this writing, it is OPEC's single largest competitor.

Weekly reports

There are two reports released each week that provide fundamental information on the domestic oil picture. Unless they are delayed by a holiday, the figures released by the API and the DOE come out on Wednesday morning. The API collects information from its members on a voluntary basis. The DOE collects figures from oil companies on a mandatory basis. Because of the mandatory reporting requirement, the DOE numbers are generally considered more accurate.

First tier of sophistication

In both reports, traders look first and foremost at the week-to-week change for inventories of distillate, gasoline, crude oil, and in refinery utilization. Those four figures are forecast ahead of the report by wire services, which poll analysts and traders. Everybody looks at these figures, including die-hard technical or chart traders. Big changes in the week-to-week inventory figures can create knee-jerk reactions in prices.

Also included in this are comparisons of inventory levels against year-ago levels or against multiyear averages.

Second tier of sophistication

The second degree of detail looked at commonly includes imports of distillate, gasoline and crude oil, crude oil runs, and domestic production of distillate, gasoline, and crude oil. These figures give an implied demand for the two refined products. The reported change in crude oil stocks can also be compared to the difference between the combination of domestic crude oil production with imports and crude oil runs.

To calculate this, the following equation can be used:

$$\text{Actual build or drawdown in crude oil stocks} = (\text{Domestic crude production} + \text{crude oil imports} - \text{crude oil runs}) \times 7$$

In reality, these are never equal, and many traders will keep a running count going of whether barrels are owed to the report by the market or are owed to the market by the report.

To compute implied daily consumption of gasoline or distillate (heating oil/diesel):

$$\text{Daily consumptions} = (\text{Daily production} + \text{daily imports}) - (\text{stock change}/7)$$

Many analysts and traders will compute multiweek averages of supply and demand and run them through different regression or forecasting models to project future supply and demand. The relevant inputs needed to do the same thing are given in Table 9–1.

Table 9–1 Computing multiweek averages of supply and demand

Commodity	Supply Inputs	Demand Inputs
Crude Oil	Imports, Domestic Production	Crude Oil Runs
Distillate	Production and Imports	[(Prod + Imp) − (Stock Change/7)]
Gasoline	Production and Imports	[(Prod + Imp) − (Stock Change/7)]

Clearly, these stock levels should be compared to those of the previous week and the prior year, or perhaps even to an average of a number of past years.

Third tier of sophistication

Traders like to look at the inventory situation by Petroleum Administration for Defense Districts (PADDs). In World War II, each district received a certain number of rationed barrels, depending upon past usage patterns. The war is over, but the PADDs remain. The districts are as follows (see Table 9–2):

Table 9–2 PADDs

PADD I	East Coast States
PADD II	Midwest States
PADD III	Gulf States
PADD IV	Rocky Mountain States
PADD V	West Coast States

If there is a particularly large increase or decrease in crude oil stocks, traders will break them down into PADDs to see more specifically what is going on. Any large or unusual changes in PADD V crude oil inventories are generally discounted because barrels routinely appear and disappear on the West Coast. It is crude oil's Bermuda Triangle. Big changes in PADD III represent additions or subtractions from stocks held by the main refining district in the country. If there is a large buildup in PADD III, it could mean that there was a unit unexpectedly idled and not reported. It could also mean that refiners plan to increase runs (the amount of crude oil processed into refined products) in the near future.

In the winter, traders will take a look at PADD I distillate stocks and will break down stocks into high sulfur and low sulfur distillate stocks. These are intended for use as heating oil or diesel, respectively.

Traders will also look at the breakdown between reformulated (RFG) and conventional gasoline stocks. They will also look at blending stocks, RBOB (reformulated gasoline blendstock for oxygenate blending), and ethanol levels.

Comparing historical stock levels

In the mid-1990s, the U.S. refining industry moved to the *just-in-time* inventory system. That means that inventories held before fall 1994 are not as meaningful in comparison with more recent years. There have been times when the just-in-time inventory system was not in time, and terminals have run out of jet fuel, heating oil, diesel, or gasoline for a day or longer over the last few years. Despite this major failing, vociferous shareholders insist on squeezing every fraction of a penny from refiners, and the system is here to stay.

WEEKLY "COMMITMENTS OF TRADERS" REPORTS

The Commodity Futures Trading Commission (CFTC) releases "Commitments of Traders" reports every week on Friday afternoon. These break down the open interest into three categories (see Table 9–3). There is *long or short only,* or large speculators, which is for large traders with reportable positions. These are mostly funds. There is the *nonreportable* category, which is comprised of small hedgers and speculators who do not have enough contracts to constitute a reportable position. And then there is the *commercial* category, which consists of large concerns that actually produce or use the commodity.

Table 9–3 "Commitments of Traders" reports

Category	Players Included
Long or short only	Large speculators, funds, big locals
Nonreportable	Small speculators, heating oil distributors, gatherers, resellers, smaller locals, smaller fleet operators
Commercial	Refiners, producers, pipelines, bigger cargo traders

How to analyze these reports

This same theme will be covered elsewhere in the book. But the usefulness of the commitments reports is that they give one a breakdown of the relative proportion of commercial positions to fund positions and smaller speculative and hedge positions.

To make a long story short, at the beginning of a move the trade is likely to hold a disproportionate number of the positions that are about to become more profitable. The funds will hold a disproportionate number of the positions that are about to have the market turn against them. The market turns, and funds get rid of their losing positions and jump on board the new trend. The trade takes profits on its improving positions on a scaled-up or scaled-down basis. When the move passes its midpoint, the funds will be seen adding fresh positions in the new direction, while the trade will take on fresh scaled-up shorts or scaled-down longs. This process continues until the new trend becomes an old trend, and then the process repeats.

Who does better as a group, the funds or the trade? It is hard to say. The early funds do well and the late commercial accounts do well. But neither category seems to have any real advantage as a group, per se.

The funds are the aggressive group that takes the action to the trade, but the trade is often the defensive one that eventually breaks the charge and puts it to flight.

Sometimes there is a steady increase in the net number of longs in the *long or short only* category, and longs outnumber shorts by a substantial ratio. When this occurs, the market is poised for a sharp sell-off. In addition, the trade could be net short by a large number or ratio (commercial accounts hold much bigger positions but turn them over much less frequently). If this is the case, then the combination of these two positions argues heavily for a sharp sell-off.

The opposite is true when the funds are disproportionately short and the trade is significantly net long. Under those circumstances, one should be looking for a sharp rally that could begin a longer-term advance. By the time the advance is over, the funds will be holding a much larger proportion of longs than shorts, while the trade will be net short.

It is simply a matter of which group is going to trade more urgently and more often. That is decidedly the funds. Funds go to the market rapidly, while the trade moves in a more dignified manner. The trade uses limit orders on a scaled-up or scaled-down basis and lets the market come to it.

Interpreting these reports is an underdeveloped art. Experience and a little homework will sharpen any ability, but this one more than others. If open interest moves in the same direction as prices, it is supportive. If they move in opposite directions, it is generally considered bearish.

Technical Analysis

TECHNICAL VS. FUNDAMENTAL ANALYSIS

A pure technician believes that all the known fundamentals are already incorporated in the current price structure. Anything important will lead to a buy or sell order, and that will be registered in the stream of prices. There may be worthwhile fundamentals that excite the press or market analysts, but the technician will only be influenced by the factors that excite real traders into giving orders. And those always lead to price moves. There are fundamental factors that grab attention in the media, but the fundamentals matter *only to the point that they lead to new buy or sell orders*. Those orders are what move the market.

Pure technicians believe that there are fundamental factors of which they will never be aware. But they also believe that those factors will leave their footprints in the market's price activity. They may not know what is making traders buy or sell, but they can follow the effects of that buying or selling on the market.

Technicians believe that a trend in motion will stay in motion until it has run its course. They believe that the fundamental factors making traders buy today will continue to influence them into buying again tomorrow. Fundamentalists argue that charts show only yesterday's influences, and that charts show only where prices have been, not where they could be going.

This can be compared to being in a car with the windshield blacked out. The fundamentals can be followed and considered to be the fuel input for the imaginary vehicle. The speedometer can be monitored, which is the equivalent of price action or movement. It is also possible to tell how much fundamental fuel has been used and see how fast and far it has moved prices. And from there it is possible to determine the market's gradient, or trend. Fundamentalists and technicians alike agree that the trend determines everything.

In April and May 1984, prices advanced briskly every time an Iranian plane bounced an antitank missile off the quarterdeck of an Arabian tanker. The market was bullish, and prices advanced on anything more threatening than a rowboat and a BB gun. But in June, there were listing tankers with Iranian oil gushing out of hulls ripped by wave-skimming Exocet missiles, and the market dropped, no longer impressed by the tanker war. The fundamental fuel was increased, but the market slipped backwards, nonetheless. That was a sign that moving higher had become an uphill struggle, and it was a sign that prices could move lower.

The trend had changed, and the market's most bullish fears had already been discounted before they even happened. Once they were realized, a certain disappointment set in and anticipation dropped dramatically.

Most traders would feel as lost as a cork on the ocean without the combination of fundamentals and technicals. They work in concert and against each other to define the trend. It is not enough to fill up the car and believe that the fuel will take one somewhere. It is nice to know whether one is going uphill or downhill before one steps on the accelerator.

Types of Charts

Bar charts

The most commonly used chart is the bar chart. These are easily constructed. One draws a straight vertical line from the day's high to its low, and one then makes a horizontal hash-line to the right of the line to signify the day's close. A hash mark to the line's left would represent the day's opening price. Figure 10–1 shows one day's entry on a bar chart.

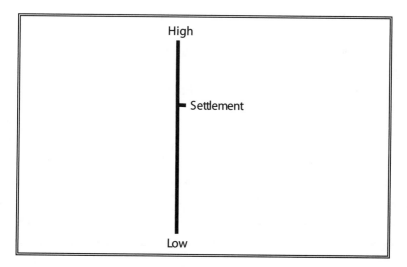

Fig. 10–1 Bar chart

There are other types of charts. Point and figure charts keep track of only price, with no axis for time. They require constant monitoring to be useful, and a whole chapter could be written on their use and interpretation. These days, only the most dedicated and advanced traders use them, such as locals on the floor. Few upstairs traders keep handmade charts any more.

When I started in this business, I used to update 300 charts each day by hand, and I needed to work weekends just to replace the ones that expired each week. Now everyone relies on machines, and I do not keep any handmade charts. And hardly anyone else I know keeps them, either.

Are they missed? Certainly they are missed, like chalkboards and tape tickers and other signs of days gone by. But they were a good way to start out. A trader got a better sense of the market's pulse by keeping charts updated by hand. A person who is still early in the learning curve should keep about 15–30 charts by hand. These could include the first three months' bar charts, selected spreads, point and figure charts, weeklies, and RSI. It is also helpful to keep a diary or journal, not hesitating to record mistakes or minor revelations. A real education in this business is too expensive to make the same mistakes twice. And there is no need to do so.

Swing charts are also helpful. A swing chart connects lows to highs and highs to lows, and it gives one a clear and uncluttered view of the market. But starting out, it is important to keep as many different kinds of charts as possible. Eventually, there will come a point where there are too many inputs to make a decision. At that point, it makes sense to switch from a shotgun to a rifle. But it is wise to start with the broader approach, as it is a faster way to gain an understanding of the market.

HOW PRICES ARE MADE

The first time on the floor of any futures exchange is an occasion. There is a great deal of yelling and pushing and shoving, and one invariably walks away wondering how it could possibly work at all. It seems like a miracle that prices get traded and recorded as accurately as they do. The proportion of out-trades, or trades not recognized by both parties as having occurred, is remarkably small given the seeming mayhem on the floor.

Prices in the futures market are created in much the same way as they are in the stock market, through a series of bids and offers that eventually converge at an agreed-upon price. But futures are traded through open outcry, rather than through specialists who know where all the buying and selling is.

Inside information is an illegal advantage in stocks. In futures there is no such thing. Any information available can be used, and sometimes it helps. The key is to get information before everyone else does. However, it is worthless unless everyone else gets it eventually. Information that never makes it to the market is useless.

The most important people in the market are those who have existing positions. Traders looking to initiate a position are relaxed. Traders looking to exit a position are much more involved. *There is no more eager buyer than a short losing money, and there is no more willing seller than a long in negative territory.* They are not relaxed at all. Futures do not dillydally, and traders with losing positions are not afforded the luxury of riding it out.

Any losses are realized whether the position is open or closed. *Every loss is real right away.* Futures always reach maturity long before those trading them do. Since positions are taken to assume or avoid risk, futures are not investments.

The only difference between speculating in futures and gambling is that speculators shoulder existing risks, while gamblers accept created risks. Those in the trade with existing risks who do not hedge or offset those risks are speculators. An unhedged exposure in the cash market is not different in any significant way from a speculative position held by an out-and-out speculator.

THE TREND

Random markets have been constructed based on the arbitrary addition or subtraction of a random number. They should end up moving sideways, but they do not, of course. They trend just like the markets do. Something in nature must compel trends. One of the most fascinating examples is the tendency of temperatures to trend. There frequently can be 9, 10, 15, or 18 months in a row with temperatures consistently warmer or cooler than normal. It takes about six weeks to change one of these trends.

The oil markets average four trends a year. Sometimes, there will be three or five. It is extremely rare to get many more or less than that.

The first book I was told to read when I started in this business was *Reminiscences of a Stockbroker* by Edwin Lefevre. This was a pen name for one of the greatest traders of all time, Jesse Livermore. Jesse was short and selling during the stock market panic of 1907, and he helped to push the market lower (he was heavily short) in the black days of October 1929. He was a genius, but unfortunately, he shot himself after declaring his life a failure. Still, the book is vital and interesting reading for anyone trying to understand this business.

There is a character named Old Man Partridge in the book who keenly listens to everyone's opinions on the market. Whenever someone advises him to sell, he smiles and says, "Well, it's a bull market, you know." That is all he knows, and it is all he really needs to know.

A person who trades every day should make it so that the first thing seen every morning is the question, "What is the trend?" It should be written in lipstick on the bathroom mirror, carved in a trader's desk, and written on palm of the trader's hand. It has to be the first and guiding thought every day. It is amazing how easy (and hazardous) it is to stray from it.

There is a natural desire to anticipate trend changes. The increasing feeling of anticipation is so overwhelming that it drives people to get out of almost every move early. It is very hard to hold on until the end. One should consider riding the markets comparable to riding a bull. The longer one holds on, the better one does. It may help one to remember that bullish news is magnified in a bull market and ignored in a bear market. A body in motion tends to stay in motion, and the same may be said of a trend.

Two rules to remember are

- Follow the trend all the way to the end.
- Prices always move farther and faster than anyone who understands the market would dare project.

They are good rules, but nobody seems to be able to follow them. It is hard enough to enter a position at the right time and price. Even if this is accomplished, it is almost certain that it will not be followed by exiting the position in an equally well-timed manner. This leads to another rule:

- Never close out a position while planning to reenter it later at a better price.

It never works out that way. The market will keep moving, and the trader can grow old waiting for the correction. The worst mistake is to feel compelled to get back in just as the trend is about to turn. Sometimes the best trades are the ones that are not made.

The more decisions a trader makes, the higher the odds are that he will make a bad decision. As stated, there are four major trends in oil markets a year. If the trader is on the right side of two and avoids getting hurt by the other two, it is a good year. But there is not a trader in the world who can trade that infrequently. Traders need to trade. This is illustrated by the following story.

It was the last day of summer, and the surfer had come to the beach for a final day of surfing. He paddled out and spent an hour in the brine. It always seemed like there was a swell in the distance, but they just were not breaking right. And he was there to surf. It would be better to ride any old wave than to sit there waiting forever. So he decided to ride the next one in, good, bad, or indifferent. He took the next wave in, and it was mediocre. He stood up and turned around, only to see a swell on the horizon. He froze for a second and then plunged into the water, swimming furiously to get back in position. The wave took shape ahead of him. He sputtered and clawed the water desperately, as the granddaddy of all waves crested just feet in front of him.

He tried to get in front of the wave, but it picked him up like a tooth-pick in a tornado, and then it threw him down in a somersault of salt, surf, and sand. He gasped for air and thrashed around till his arm hit bottom, then he pushed his face up into the air and breathed again. The raw power of the undertow pulled him back, but he dug his hands into the sand and pulled himself away.

The surfer illustrates an important point about markets. There are only so many good trends. A trader who becomes distracted by the mediocre trends will be in a poor position to trade the really good ones and could even be hurt by them. He should not let the fact that he is there to trade become more important than following the trends.

Overtrading is the single biggest reason that people lose money in futures. Even good and talented traders succumb to the feeling that they are there to trade, every day. Good traders do not go to work to trade. They go to work to make money. And the best way to do that is to start each day thinking about the trend.

What is one looking for?

A trader is looking for signs that the tide has changed. The trend is a changing entity, but the more experienced one is at recognizing it, the better. It is best not to get too cute with the trend. When one tries to fine-tune it, one is basically second-guessing it, and that does not play well in a trending market. The trend is like the defendant in most U.S. court cases. The trader has to prove *beyond a shadow of a doubt* that the trend is over. Otherwise, it is free to return to its old stomping grounds, and it usually will. It takes time for trends to change, and they hold on tenaciously.

One should not be too proud to scrawl,

"What is the trend?"
in borrowed lipstick on one's bathroom mirror.

The key to following the trend is to doubt everything that seems to derail it. For some reason the market seems to score rounds the same way they do in boxing. The champ always wins unless there is an undeniable knockout. And the trend seems to find a way to continue until there is a technical knockout, chart-wise. Consequently, a trader should stick with the trend until he is certain it is over, and even then, he should wait for confirmation.

MARKET PHASES

The market is always in one of four phases. These phases are illustrated in Figure 10–2.

- **Accumulation.** This is the consolidation period at a bottom. At this stage, strong hands are accumulating long positions in anticipation of a move higher. The number of individuals holding long positions drops, while the average size of the long positions increases. Long positions become concentrated in fewer but stronger hands.

- **Markup.** This is the actual move higher. Prices are *marked up*.

- **Distribution.** This is where big traders sell their large holdings to smaller traders in smaller parcels. More individuals become long, but their average position becomes smaller. Long positions are moved to weaker hands.

- **Markdown.** This is the actual move lower. Prices are *marked down*.

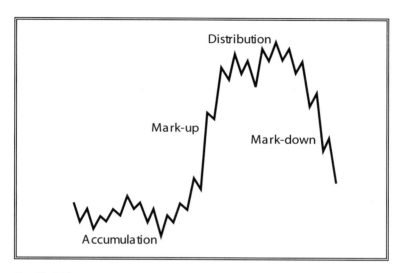

Fig. 10–2 Phases

This brings us to another rule:

• The market will always move in the direction that will hurt the most people.

That may not sound right in a zero-sum game. After all, there are the same number of longs as shorts, right? Actually, there is a difference. The number of contracts long does equal the number of contracts short, but the number of individuals holding these varies. The market will always move against the larger number of individuals holding smaller average positions.

To illustrate this point, consider an example in which the open interest is 100 contracts. One individual is short 100 lots, and on the other side, 100 individuals are each long 1 lot each. Which way will prices go? They will go lower. The 100 longs will compete with each other to sell.

If there is a panic, there will be dozens of voices to sell and possibly only 1 to buy. It does not matter that the volumes may be the same. If there are 100 people looking to sell 1 lot each, and 1 individual looking to buy 100 lots, the market should be balanced, at least on paper. But in the ring, it will not come out that way. There will be 100 mouths screaming offers and only 1 bidding. Unless the trader who is the big short is an idiot or is strict about holding his brokers to limits, he should be able to cover very favorably on a *scaled-down* basis.

The 100 sellers will compete with each other to be the lowest possible sellers. If there is only 1 big buyer, he can parcel out the buying at progressively lower prices. The sellers will compete with each other to sell at the buyer's price—his bid.

It is important to keep the voice factor in mind when considering the regular session. It is not a factor on Access, but in the ring it is. If one-half of the traders in the ring are buying and one-half are selling, then prices should stay relatively stable. But if one-fourth of the traders change their minds, there is suddenly a 3:1 disparity in the ratio of bids and offers. And that can create a mood. If there is another shift of only 5%, the ratio goes to 4:1, and a shift of 15% can push the ratio to 10:1. Consequently, small shifts in sentiment can have huge effects on prices.

At market bottoms, there are more voices to sell than there are to buy. The buyers take bigger chunks as they accumulate positions. At tops, the voices are all buying, but the sellers are dividing their sales into bite-size pieces for the smaller buyers. The big longs are distributing their long positions to many smaller traders with shallower pockets (at the top, during distribution phases).

In the 1890s, there were huge stock pools that bought shares. The big traders in these pools would buy when prices were weak, and then they would parcel out their positions in the markup and distribution phases. Once they had gotten out of their holdings, they would no longer support prices by buying stock. Prices would drop during the markdown phase until they reached a new area of accumulation where big traders would buy new positions. They would buy these with the intention of helping prices move higher, where they could sell at a profit.

SUPPORT AND RESISTANCE

Prices have a tendency to stop at previous highs or lows in their travels. There are two theories as to why. The first, and more commonly accepted, is that areas of consolidation give traders a chance to get even. Perhaps a trader sells crude oil at $22.50/bbl. The market rallies to $23.00/bbl, and the trader is out 50¢/bbl. If prices pull back to $22.50/bbl, the trader will be thrilled to get out at breakeven. As a result, that area becomes support. That is one theory.

The other theory is that prices have certain sympathetic vibrations that determine where highs and lows will occur. Although it may be tempting to dismiss this theory out of hand, there have been numerous examples of prices acting as *support* or *resistance* over very long periods. These periods were too long for anyone to still have a losing position to liquidate or cover. There are also geometrical arguments for this theory.

Whatever the reason, support and resistance levels exist. There are two important features that one must comprehend. First, if support, which is always *under* the market, is broken, it becomes resistance. By the same token, if prices break over resistance, which is always *over* the market, it automatically becomes support. Certain numbers serve alternately as important support and then resistance levels in the oil markets. The second feature is that *stops* are almost always placed just over resistance or just under support. Figure 10–3 shows support and resistance on a chart.

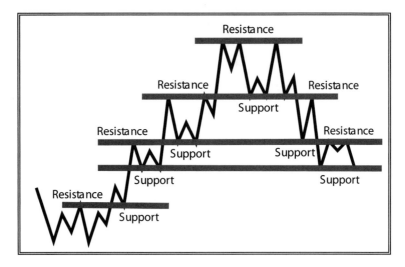

Fig. 10–3 Support and resistance

In the second week of July 1995, August heating oil prices registered highs of 47.30, 47.60, 47.50, 47.45, and 47.55. It was clear that there was good selling from 47.30 to 47.60. Prices sold off after failing to break the resistance there, falling to 45.75 and touching off sell-stops underneath the support at 46.00. On July 26, after rallying for a couple days, prices broke 47.60, and the buy-stops and technical buying over that level took prices all the way to 48.60. Resistance levels provide selling, but if they are broken, they trigger buying. Support levels provide buying to stop declines, but if the drop is powerful enough to break the support, sell-stops will be triggered.

Support and resistance are easy to identify. Traders start by trying to locate zones rather than specific numbers that have stopped previous moves. They look for groupings of numbers that have acted as breakwaters in previous advances and declines. It does not matter whether the number is a low point or a high point. If it has acted to turn prices back before, it is significant now.

*Stop orders **bring in automatic buying on a rise or selling on a drop. If a trader has a sell-stop at 45.95, the moment that price is printed on the screen, his broker is obligated to sell the specified number of contracts at the market, or at the best price available at the time. Stops bring in urgent and furious buying or selling that can move prices significantly.***

Support and resistance are the same things. They just have different names that are defined by their relationship to the current market price. Sometimes a single price will stand out as the most significant number at which prices will break. Other times, there will be small ranges that will act as support or resistance, such as the 47.30–47.60 zone described above.

Tops and Bottoms

It can be helpful to view prices as the individual drops in a river headed towards the sea. Prices, like water droplets, tend to find their level. Another helpful illustration is to imagine that a trader is stranded on a strange beach with nothing to reveal whether the tide is coming in or out other than the watermarks on the beach. And that is what prices are: watermarks on the sand.

In this example, the trader on the beach must identify as soon as possible when the tide is coming in and going out. He could guess, which is the equivalent of randomly picking tops or bottoms. However, it is much better for him to wait until the water gives positive signs that the tide has changed. If the trader actually monitors the progress of the watermarks, he eventually will be left with a watermark that is the farthest point the waves have touched. From this point he can predict when and in what direction the tide will move.

These high marks and low marks can be compared to price levels in the energy markets. Figure 10–4 shows some common tops and bottoms.

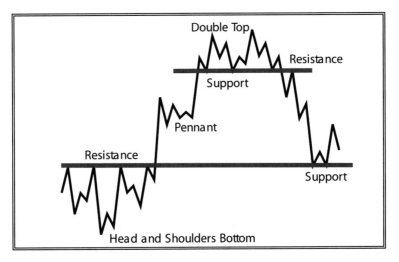

Fig. 10–4 Tops and bottoms

The most popular patterns for tops and bottoms are double tops or bottoms and head and shoulders tops and bottoms. In either case the market makes a high or a low and then challenges it. In a head and shoulders formation, prices go on to make a new high or low, but then they fail on the next attempt. In a double bottom or top, prices challenge the old high or low, but they do not move significantly beyond it.

In relation to the imaginary beach, this is a situation in which the tide is coming in, and there is a high watermark on the beach. It is not surpassed for a while, but then it finally is. Time goes by and there is another attempt to pass the high watermark, but it fails. That is the equivalent of a head and shoulders top. One must look at the undertow. It is the equivalent of the neckline on a head and shoulders formation.

Once it seems that the watermarks have gone as high as they are going to, the sure sign that the tide has turned comes when the undertow line moves back. That is the equivalent of the neckline being broken. Once the neckline is broken on a head and shoulders or double top or bottom formation, a good rule of thumb applies. In these situations prices generally will move in the new direction as far as the distance from the top or bottom to the neckline.

Tops and bottoms are not mystical patterns like constellations or the lines on one's hands. They are the physical manifestations of a trend that has turned. Furthermore, tops and bottoms take time to develop. In order to believe that a trend has changed, it normally requires enough proof to be convincing beyond any doubt. If there is any doubt, the trend is likely to reassert itself. It is generally dangerous or irresponsible to jump to conclusions without proof positive in either case.

At tops or bottoms, there likely will be either very heavy or very light volume. The characteristics of the market start to change. If moves have been quick and forceful, they may become slow or anemic. A market that has consistently closed near the top of its daily range may start to finish in the lower half of its range. There will be signs. It is important to stand back and observe the bigger picture in an attempt to decipher the signs.

PATTERNS

There are several common patterns that are seen repeatedly at major tops and bottoms. They are all derivations of the same concept: finding support in a declining market or resistance in an advancing market. If the support or resistance is strong enough, it will repel attempts to test or break it. And those failed attempts create patterns that can be recognized.

Rounded top or bottom

Energy prices occasionally make rounded bottoms and less frequently make rounded tops. They are good indicators of developing support or resistance, and they generally precede large and sustained moves. Figure 10–5 shows a rounded bottom chart.

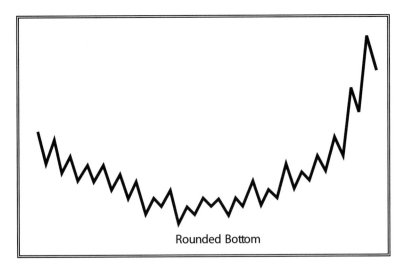

Rounded Bottom

Fig. 10–5 Rounded bottom

Key and island reversals

These formations are most frequently associated with intermediate tops or bottoms rather than final or major highs or lows. For some reason, they generally are broken in the oil complex, often after a 10-day or 2-week correction. They work best in pork bellies. Key and island reversal still are good for a brief change of trend, even if they eventually do not hold. Figure 10–6 shows three types of reversals.

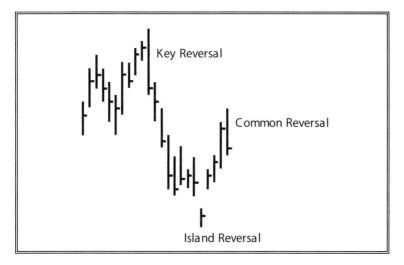

Fig. 10–6 Reversals

A key reversal day in an advancing market is one with a higher high but a close beneath the previous day's low. In a declining market, prices make a low that is lower than the previous day's low and then settle above that day's high. In order to be a true key reversal, the activity has to be on high volume.

An island reversal involves two gaps that leave one day all alone on the charts. The first gap is a kind of exhaustion gap, and the second one is a kind of breakaway gap that starts the new trend. These formations are not often seen at absolute highs or lows in energy markets. They are seen more frequently at intermediate highs or lows.

Double tops and bottoms

Double tops and bottoms are reliable trend-changing formations. They signify a failed attempt to break to fresh highs or lows in a trending market. Technical traders will usually jump all over a failed attempt to break significant highs or lows.

In Figure 10–7, the second high is just slightly over the first high. That seems to be a frequent occurrence. Locals trigger stops just beyond the previous high or low, but then they suddenly find themselves with nowhere to go. Figure 10–7 shows a double top followed by a double bottom.

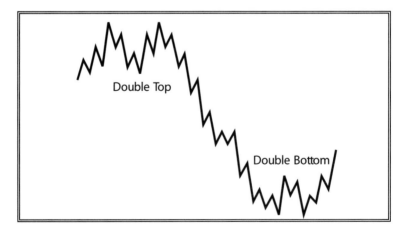

Fig. 10–7 Double tops and bottoms

Head and shoulders

Head and shoulders tops and bottoms are variations of double tops and bottoms. The first shoulder is followed by the head, which is another new high or low in the existing trend. The second shoulder is a failed attempt to break the level set by the head. It is a test of the high or low that has failed. Double tops or bottoms and head and shoulders tops and bottoms are seen frequently in energy markets. Energy markets often test highs or lows before turning. Figure 10–8 shows a head and shoulders top followed by a head and shoulders bottom.

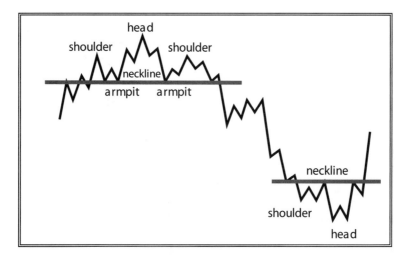

Fig. 10–8 Head and shoulders top and bottom

Traders frequently draw a neckline from *armpit* to *armpit* on a head and shoulders formation. They measure the vertical distance from the neckline (the line between the armpits) to the top or bottom of the head. They then project that same distance (in the opposite direction) from the neckline to establish an objective in the new direction.

V bottoms and tops

Figure 10–9 shows a V bottom. These represent sharp changes in market sentiment.

Energy prices can form V tops and bottoms, and sometimes these formations will be accompanied by key reversals or island reversals. Such V tops and bottoms are indicative of a sudden change in market sentiment, and they are usually created by climax buying or selling into a top or bottom. They are seen most frequently in energy markets when OPEC changes direction suddenly, when a war is suddenly declared, or when the weather changes unexpectedly.

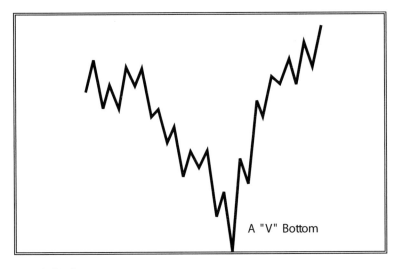

A "V" Bottom

Fig. 10–9 V bottom

Flags and pennants

Pennants and flags are continuation patterns, for the most part. They are also good patterns from which to establish objectives. One measures from the start of the move to the high or low that starts the pennant or flag and then projects that distance up or down from the bottom or top of the formation.

As an example, perhaps heating oil moves from 46.00¢ to 48.00¢ (Leg A), and then forms a flag with its lower end at 47.50¢. One could measure the first move, 2¢, and then add it to the lower level established in the flag formation, in this case 47.50¢. This gives an objective of 49.50¢ (Leg B). Figure 10–10 illustrates flags and pennants. In flags or pennants, Leg A is usually equal to Leg B.

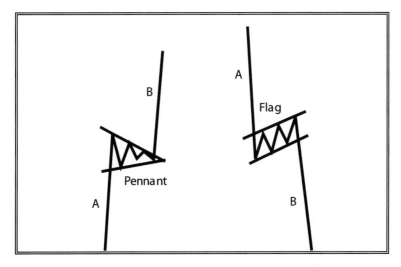

Fig. 10–10 Flag and pennant

Gaps

Three types of gaps are illustrated in Figure 10–11. The first is a break-away gap, defined as the initial penetration of a consolidation. Prices usually follow a breakaway gap. It is a very reliable signal.

The second gap illustrated is a common gap, or one that does not have that much significance. Some traders believe that gap areas will always be filled. However, there does not seem to be a consistent correlation. Common gaps do act as good support or resistance levels.

The third type of gap illustrated is an exhaustion gap. These occur when the market is near an important high or low, and they are usually a sign of final short covering into a high or long liquidation into a low. They occur when traders throw in the towel, exiting all positions.

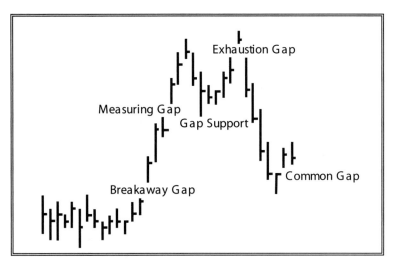

Fig. 10–11 Gaps

Trendlines

Many technicians use trendlines to help them trade the market. Here again, the bottom-line is identifying and sticking with the trend. When prices approach the trendline, traders will buy and place a stop underneath the line. If the line is broken, particularly with a close underneath it, these traders will sell at the market. Figure 10–12 illustrates a trendline.

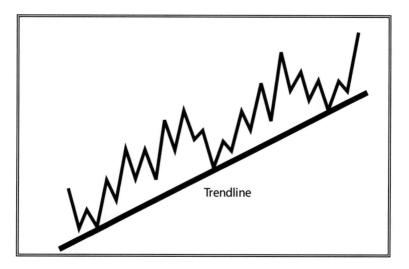

Fig. 10–12 Trendline

Moving averages

Many traders use the moving averages of a commodity to help them define the existing trend. The application is fairly straightforward. If the moving average of a fewer number of days is higher than the moving average of a larger number of days, then the trend is higher. If the average of fewer days is below the average of more days, the trend is lower. There are numerous refinements that can be added, including comparisons of the averages against themselves on a day-by-day basis. They can be used to compute overbought or oversold pressures and momentum. But the unifying theme of all moving average analysis is the comparison between the average of a fewer number of days with the average of a larger number of days.

Moving averages will often catch major and intermediate trends. But they frequently give false signals over the shorter term. The key is to select averages that will catch moves early enough without giving a lot of false signals. That is the hardest part of using averages to determine one's positions: selecting the right timeframes, the optimal number of days to compare.

For the oil complex, the 2-day, 3-day, and 8-day averages work pretty well. Longer durations do not catch new moves quickly enough. Once a trend has changed, it remains that way until all of the averages are pointing in the opposite direction. In addition, the latest day's settlement is helpful as well. In other words, if the 2-day has been over the 3-day, and both have been over the 8-day, it would be considered an uptrend. That would remain unchanged until the next time that the day's settling price is under all three averages. In other words, the 2-day average would be beneath the 3-day and 8-day averages, and the 3-day average would be under the 8-day average. At that point, the trend would be changed to lower.

In order to use averages successfully, other inputs need to be ignored, and the relationship of the averages must be strictly followed. It is difficult using averages to catch an absolute high or low, but this approach will never miss a big move, either. Traders will get chopped up in sideways markets but will make it back in trending markets.

Moving averages are hard to use in conjunction with other indicators. Most traders who use them successfully either follow them religiously or hardly at all.

CONTRARIAN INDICATORS

"The last time the majority was right, it elected Teddy Roosevelt—and even that it could not do twice." —A contrarian curmudgeon

That is the basic assumption behind contrarian analysis. It is based on a belief that the majority of traders will be wrong for all the right reasons as they embrace factors that have already been discounted. Contrarians believe that the majority is almost always wrong. A contrarian will never be comfortable when too many other traders agree with him. When people accept something as true, the contrarian will ask why everyone is not wealthy. He will respond that the world just is not set up to let everyone be rich at the same time.

The underlying theory behind contrarian analysis is that traders put their money down before they are willing to tell reporters what they plan to do. However, there are not many situations in which traders tell reporters they are bullish on a Monday with the intention of buying on the following Tuesday. It is far more likely that the trader will buy on

Joseph Kennedy made his money in bootlegging, but he increased its value by selling out his stock positions two weeks before the 1929 collapse. What tipped him off? His shoeshine boy was touting stocks. If the buying had gotten that far down the ladder, he reasoned, there was not much left. He sold out that day.

Monday and explain why it is still the thing to do on Tuesday. Of course, once they have bought the market, the only thing left for them to do is to sell it. They would like to sell it higher, but sooner or later, they will sell it. When there are too many longs looking to sell at higher prices, the selling will come before the higher prices.

Sometimes there is a front-page story in one of the most noted newspapers, or a lead story on network television, about a commodity price being high or low. When this occurs, that commodity usually will change direction within a few days. This has an astounding correlation rate of better than 90%. It is the most reliable indicator as well as a classic contrarian indicator.

Overbought and oversold

Prices get overbought and oversold as they move aggressively higher or lower. A market that is overbought or oversold is like a rubber band stretching. Sooner or later, the rubber band has to snap back.

When the market is stretched too far in one direction or another, it can often be a sign that prices are getting ready to reverse direction. There are many different measures of overbought and oversold pressures. They can be as simple as a simple deviation from a moving average of a few days. They can be complicated numbers determined by complex formulae. The bottom-line is simple. They are trying to measure overbought or oversold pressures against historical levels.

RELATIVE STRENGTH INDEX (RSI)

The RSI is an extremely popular oscillator developed by Welles Wilder. It is an indicator of overbought or oversold pressures as well as being a measure of momentum and velocity.

When the RSI gets up to 80% or down to 20%, it is generally a sign that prices are near a major top or bottom, respectively. Different contracts vary, but as a general rule, it holds. The RSI also is a good indicator of market momentum. If crude oil settles today at $16.50 with an RSI of 35%, after touching $17.00 two weeks ago with an RSI of 20%, then it is stronger today at $16.50 than it was two weeks ago at $17.00. That is called positive divergence and is considered bullish. If it had an RSI of 20% today and settled at $17.00, after trading at $16.50 two weeks ago with an RSI of 35%, then it would have negative divergence, which is considered bearish.

The role of the media

Reporters need to explain why everything happens the way it does in these markets. If prices advance, they need to know why prices moved higher, and they are not generally interested in theories about how the trend is about to change. The Friday before Black Monday in the stock market, all the press reports were bullish. On the following Tuesday, the tone had changed dramatically. There had been warning signs all along. Fundamental weaknesses had been glossed over, and the market was overdue for a correction. The press is generally only interested in explaining the most recent move. So, when prices advance, they will highlight the bullish factors, and when the market weakens, they will focus on bearish factors.

As noted previously, probably the most reliable signal that a market is getting ready to turn comes from the press. Front-page stories in major newspapers or journals and lead stories on the nightly news typically signal the end of the trend. Any time a commodity makes it as lead story in any of these media, it has gone through a process that almost ensures that the trend is about to change.

The scenario begins with a reporter calling when oil prices are low. Most of the brokers, traders, and analysts are bearish. So the bearish story is buried in the back of the newspaper. Prices start to advance. There now may be 1 bullish source, but the other 9 out of 10 are still thinking about the bearish possibilities, so the story appears deep in the second section. Prices continue moving higher, and now there are 3 sources thinking higher. The bullish parts of the story develop some, and the story moves closer to the front of the newspaper.

Prices keep getting stronger, and eventually 9 out of 10 market observers are bullish. The story comes together, and there is plenty of supporting evidence bolstered by consensual quotations. It is time for a lead story. All the bullish news is out, and reporters need to explain why prices have moved higher. Analysts are waiting in line to explain why.

Precisely when the window-shoppers know *exactly why* prices are going to move higher, those prices almost certainly will not do so. Once the little old ladies with buckets of quarters are ready to invest, the move is sure to be in trouble. There will be front-page stories on a commodity in major newspapers or magazines and the networks may use it as the lead story. When this occurs, the market in question reliably will change its trend within 10 days of the story.

OPEN INTEREST AND VOLUME

The accepted wisdom is that any move accompanied by heavy volume should be given greater respect, which is true to an extent. Very heavy volume tends to occur at tops and bottoms, and it slips during the actual markup or markdown phase. Heavy volume at a high or low, followed by lighter volume on a movement in the opposite direction of the existing trend, are good signs that the market is about to change direction.

If there are two days of equal price movement in opposite directions, the one with the higher volume is generally considered a better harbinger of the market's direction. Interpreting volume is much more of an art than a science. There is no substitute for experience in reading volume.

Open interest is interpreted in very straightforward manner. If the open interest moves in the same direction as prices (i.e., lower or higher), then it is considered bullish. If the two move in opposite directions, it is considered bearish. This can be explained by considering that if prices are higher on increasing opening interest, it signifies new buying that observers can assume will continue. If prices drop and open interest falls, it is a sign of long liquidation (or traders getting out). It is considered a good sign that the market is moving into stronger hands, or away from weaker hands.

If prices are higher, but open interest is lower, it is considered a sign of short covering, or brief but motivated buying. Once the shorts have covered, prices should be expected to return to lower levels until fresh buyers will again be interested. If prices are lower and open interest is higher, it is considered a sign that fresh selling will persist. This often occurs at higher prices that commercial traders find irresistible. Sometimes, though, the new selling can come from trading funds.

"COMMITMENTS OF TRADERS" REPORTS

Every week, the CFTC releases a breakdown of the markets' open interest, or outstanding positions, by categories. This is useful information that can help identify who holds what. The categories give a clear view of the relative number of contracts held by funds, small speculators and hedgers, and the trade (commercial accounts).

Funds are in the markets to make money, and their trades are governed by factors that can change very quickly. If the technical outlook or something in the news changes these large traders' perception of the market, they can react quickly with spectacular results. If too many small speculators and hedgers are in a market, there can be a sudden stampede of smaller orders that will still require a great deal of lung power on the floor.

The commercial trade has positions because it wants to buy or sell the physical product. Commercial traders have positions on for real-world reasons, and they are less likely to be swayed by day-to-day changes or rumors. As a result, the market tends to move in favor of the trade simply because it changes its positions less frequently. It is a matter of turnover. Speculators change their minds more frequently than commercial traders with wet barrel needs.

As a market trends, it tends to gain speculators on the short side in a downmove or on the long side in an upmove. The trade tends to increase its long positions on declines and its short positions on advances. When there are too many speculative shorts and the trade is heavily long, prices tend to rally. When there are too many speculative longs and the trade is on the short side, prices tend to sell off. And yet, speculative buying or selling can press prices higher or lower for weeks on end.

Another ratio to watch is that of large speculative longs to shorts. When they reach disproportionate levels, a change should be expected. As long as open interest increases on a move (until reaching that ridiculous stage), the move is expected to continue. It is important to watch open interest and volume on a daily basis and then try to integrate them into the larger technical picture.

11
Seasonal Tendencies and Market Psychology

HEATING OIL

The March seasonal period is one of the most exciting seasonal periods in the oil complex. The seasonal tendency for oil prices to reach a significant bottom in the first half of March is well established. February 2003 lacked the usual decline in prices because traders were apprehensive of the potential effects of a war with Iraq. Typically February is the weakest month on the energy calendar.

There is a great deal of research on heating oil futures, because they have traded for the longest period of time. This is considered primarily a heating oil seasonal tendency, although it works even better in gasoline. Trading strategies and tactics that have worked with this seasonal tendency will be examined in the following examples.

Table 11–1 looks at the settlement prices for June heating oil over the last 24 years. A check indicates that the seasonal worked, an equals sign that it broke roughly even, and an x indicates that it failed.

Table 11–1 June heating oil futures

Year	3/1	4/1	4/15	5/1	5/15	Result
2003	86.28	72.45	72.88	69.10	75.04	✗
2002	59.10	68.94	63.84	67.42	67.79	✓
2001	69.89	67.01	79.23	76.47	76.63	✓
2000	71.57	65.85	62.93	67.44	77.25	✓
1999	33.05	43.40	43.08	44.47	42.95	✓
1998	42.92	43.83	44.28	45.78	41.64	✓
1997	52.99	53.95	53.89	54.33	56.48	✓
1996	53.42	57.08	62.62	54.16	54.81	✓
1995	46.44	47.47	48.79	51.09	49.86	✓
1994	45.55	45.80	47.40	47.57	47.81	✓
1993	55.79	56.25	55.83	55.88	54.08	=
1992	50.12	53.97	55.00	56.70	57.18	✓
1991	50.49	52.18	56.32	54.27	55.52	✓
1990	53.56	54.51	51.19	52.97	51.72	=
1989	46.94	50.62	52.25	49.46	49.87	✓
1988	41.33	44.84	48.44	45.95	47.98	✓
1987	43.33	48.92	47.44	48.15	51.89	✓
1986	38.33	35.33	37.66	40.56	43.62	✓
1985	68.40	74.50	73.90	71.89	70.32	✓
1984	77.20	77.31	78.01	78.63	81.37	✓
1983	70.24	75.81	77.60	78.33	76.68	✓
1982	75.34	77.69	86.98	89.32	92.90	✓
1981	96.95	95.10	96.50	91.49	92.46	✗
1980	78.25	77.00	77.85	78.00	78.30	=

Settlement on day closest to day listed.
Prices in ¢/gal.

TRADING STRATEGY

The rule calls for the purchase of June heating oil in the first two weeks of March. At some point between then and mid-May, there will be a profit. Some years it is better to sell in April than in mid-May, but many times, it makes sense to hold the long positions until then.

Cameron Hanover likes to institute a *time-buying* plan whereby a certain number of lots is divided up over the trading days in the first half of March. If there are 10 trading days and a client needs to buy 10 lots, the plan would be to buy 1 lot a day, just like a multiple vitamin. If there are 50 lots to buy in the same trading period, the plan would be to buy 1 lot each hour. Once profitable prices are reached in April, the client can start unwinding them, using overbought statistics to help adjust the timing and number of each sale.

The biggest question with this trade is when to get out. There is not as hard-and-fast a rule for getting out as there is with getting in. Furthermore, the system has never really worked with a stop-loss point, so there are no guidelines on avoiding disastrous losses other than the rules used in normal trading. This needs to be viewed as a long-term trade that necessitates a commitment to stay with it year after year.

In the 24 years listed in Table 11–1, this trade was a big loser in just 2 years. It broke even in 3 years (although there were opportunities to get out at a profit), and it *could* have made money in the other 19 years. That is 8.33% wrong, 12.50% undecided, and 79.17% "right." The quotes around "right" are there because client did need luck and timing in 2 of the last 7 years to make it work. Still, it yielded profits for good and/or lucky traders in those years.

Most years offer steady gains that more than offset the break-even years or the worst possible loss in the couple of years that did not work: 1981 and 2003. But since there was no history of any tendency in the 2½-year-old heating oil contract in 1981, there was no reason to have made that trade in the first place. Again in 2003, there was no reason to make this trade. The United States was just about to invade Iraq, and prices were high on war fears in February. It was the only year that we did not try to use this marketing strategy since we started following this seasonal. The war skewed the seasonal tendency.

Tactical variations

There are a few different variations on the basic theme, and they tend to make the trade more disciplined for those who are unhappy with the rather vague procedure outlined above.

Short-term trade. This plan calls for the purchase of the full line on March 1, and the sale of one-half of the position on April 1 and the other half on April 15. Table 11–2 illustrates this point.

Table 11–2 June heating oil futures—short-term trade

Year	3/1	4/1	4/15	Average	P/L	Result
2003	86.28	72.45	72.88	72.67	-13.61	✗
2002	59.10	68.94	63.84	66.39	+7.29	✓
2001	69.89	67.01	79.23	73.12	+3.23	✓
2000	71.57	65.85	62.93	64.39	-7.18	✗
1999	33.05	43.40	43.08	43.24	+10.19	✓
1998	42.92	43.83	44.28	44.05	+1.13	✓
1997	52.99	53.95	53.89	53.92	+0.93	✓
1996	53.42	57.08	62.62	59.85	+6.43	✓
1995	46.44	47.47	48.79	48.13	+1.69	✓
1994	45.55	45.80	47.40	46.60	+1.05	✓
1993	55.79	56.25	55.83	56.04	+0.25	✓
1992	50.12	53.97	55.00	54.48	+4.36	✓
1991	50.49	52.18	56.32	54.25	+3.76	✓
1990	53.56	54.51	51.19	52.85	-0.71	✗
1989	46.94	50.62	52.25	51.43	+4.49	✓
1988	41.33	44.84	48.44	46.64	+5.31	✓
1987	43.33	48.92	47.44	48.18	+4.85	✓
1986	38.33	35.33	37.66	36.49	-1.84	✗
1985	68.40	74.50	73.90	74.20	+5.80	✓
1984	77.20	77.31	78.01	77.66	+0.46	✓
1983	70.24	75.81	77.60	76.70	+6.46	✓
1982	75.34	77.69	86.98	82.33	+6.99	✓
1981	96.95	95.10	96.50	95.80	-1.15	✗
1980	78.25	77.00	77.85	77.42	-0.83	✗

Settlement on day closest to day listed.
Prices in ¢/gal.

In 18 out of 24 years, there was a profit in this trade. The remaining 6 years returned a loss, assuming fills at the day's settlement levels. Following this system to the letter would have resulted in net profits of 49.35¢/gal, or $20,727.00 per contract before commissions and fees. That is an annual average of 2.06¢/gal. The biggest losses were 13.61¢/gal in 2003 and 7.18¢/gal in 2000. The largest gain was 10.19¢/gal in 1999. Prior to those years, the biggest loss was 1.84¢/gal in 1986, and the biggest gain was 6.99¢/gal in 1982. The violent trading years of 1999, 2000, and 2003 increased the volatility in this trade substantially.

We did not recommend making this trade in 2003, because prices had just started coming down from all-time highs of $1.31/gal in heating oil. American forces were on the brink of removing Saddam Hussein from power. It was fairly clear at the time that the seasonal was unlikely to take precedence over the extreme political events of the moment.

Longer term trade. These trades require buying half the position on March 1, the other half on April 1, and the sale of everything on May 15. It has worked in 17 out of 24 years, and has made a net profit of 63.16¢/ gal per contract, or $26,527.20 profit before fees and commissions, per contract. It has gained an average of 2.63¢/gal per year. The biggest loss was 4.54¢/gal in 1981. The biggest winner was 12.80¢/gal in 1982.

This trade worked in three of the years that the trade above did not. It did not work in three of the years in which the first trade did. It has been a big winner in three of the last five years. It would have worked well in 1998 if we had sold it out on May 1 rather than May 15, but that gets away from rules that have been designed to inject some order into a disorderly process. The year-by-year breakdown for the trade is listed in Table 11–3.

Table 11–3 June heating oil futures—longer term trades

Year	3/1	4/1	Average	5/15	P/L	Result
2003	86.28	72.45	79.37	75.04	-4.33	✗
2002	59.10	68.69	63.90	67.79	+3.89	✓
2001	69.89	67.01	68.45	76.63	+8.18	✓
2000	71.57	65.85	68.71	77.25	+8.54	✓
1999	33.05	43.40	38.23	44.47	+6.24	✓
1998	42.92	43.83	43.38	41.64	-1.74	✗
1997	52.99	53.95	53.47	56.48	+3.01	✓
1996	53.42	57.08	55.25	54.16	-1.09	✗
1995	46.44	47.47	46.95	51.09	+4.14	✓
1994	45.55	45.80	45.68	47.57	+1.89	✓
1993	55.79	56.25	56.02	55.88	-0.14	✗
1992	50.12	53.97	52.04	56.70	+4.66	✓
1991	50.49	52.18	51.33	54.27	+2.94	✓
1990	53.56	54.51	54.04	52.97	-1.07	✗
1989	46.94	50.62	48.78	49.46	+0.68	✓
1988	41.33	44.84	43.08	45.95	+2.87	✓
1987	43.33	48.92	46.13	48.15	+2.02	✓
1986	38.33	35.33	36.83	40.56	+3.73	✓
1985	68.40	74.50	71.45	71.89	+0.44	✓
1984	77.20	77.31	77.26	78.63	+1.37	✓
1983	70.24	75.81	73.03	78.33	+8.30	✓
1982	75.34	77.69	76.52	89.32	+12.80	✓
1981	96.95	95.10	96.03	91.49	-4.54	✗
1980	78.25	77.00	77.63	78.00	+0.37	✓

Settlement on day closest to day listed.
Prices in ¢/gal.

Simple, longest term trade. This plan calls for buying everything on March 1 and selling it all on May 15. This trade worked in 19 out of the last 24 years, and it had net gains of 74.70¢/gal, or $31,374.80 per contract, before any fees or commissions. Before the war-impacted price decline of 2003, the biggest loss was 5.46¢/gal in 1981, and the biggest gain was 13.98¢/gal in 1982. A trader who made this trade, year in and year out, would have made 3.11¢/gal per year, on average. Table 11–4 shows this trade.

Table 11–4 June heating oil futures—longest term trade

Year	3/1	5/15	P/L	Result
2003	86.28	75.04	-11.24	✗
2002	59.10	67.79	+8.69	✓
2001	69.89	76.63	+6.74	✓
2000	71.57	77.25	+5.68	✓
1999	33.05	42.95	+9.90	✓
1998	42.92	41.64	-1.28	✗
1997	52.99	56.48	+3.49	✓
1996	53.42	54.16	+0.74	✓
1995	46.44	51.09	+4.65	✓
1994	45.55	47.57	+2.02	✓
1993	55.79	55.88	+0.09	✓
1992	50.12	56.70	+6.58	✓
1991	50.49	54.27	+3.78	✓
1990	53.56	52.97	-0.59	✗
1989	46.94	49.46	+2.52	✓
1988	41.33	45.95	+4.62	✓
1987	43.33	48.15	+4.82	✓
1986	38.33	40.56	+2.23	✓
1985	68.40	71.89	+3.49	✓
1984	77.20	78.63	+1.43	✓
1983	70.24	78.33	+8.07	✓
1982	75.34	89.32	+13.98	✓
1981	96.95	91.49	-5.46	✗
1980	78.25	78.00	-0.25	✗

Settlement on day closest to day listed.
Prices in ¢/gal.

GASOLINE

In unleaded regular gasoline, the tendency for prices to advance from early March is even greater than it is in heating oil. Following the same loose guideline of buying in the first two weeks of March and selling before May 15, the seasonal tendency has worked in 18 out of 19 years, or 94.74% of the time. We like that kind of historical track record. The only year it did not work was 2003, when prices ran up in February and March before the war with Iraq. Table 11–5 provides the actual figures.

Table 11–5 June unleaded regular gasoline—loose trade guidelines

Year	3/1	4/1	4/15	5/1	5/15	Result
2003	103.39	85.22	84.56	79.02	86.73	✗
2002	71.26	84.37	78.33	80.48	78.82	✓
2001	85.19	87.98	100.70	105.84	100.36	✓
2000	88.10	81.55	78.95	83.27	96.79	✓
1999	40.03	52.89	52.16	55.19	52.00	✓
1998	52.15	52.11	52.74	54.29	50.75	✓
1997	61.63	62.31	61.47	62.63	64.38	✓
1996	57.66	64.59	69.12	67.42	67.20	✓
1995	55.89	57.60	60.76	64.15	64.90	✓
1994	46.83	47.62	50.78	50.47	51.17	✓
1993	60.36	61.22	60.95	61.81	59.12	✓
1992	60.15	63.13	60.61	64.69	63.68	✓
1991	61.67	63.71	70.21	70.83	69.08	✓
1990	62.52	64.95	59.55	61.03	63.5	✓
1989	53.59	65.00	68.21	72.61	66.09	✓
1988	45.27	49.00	52.15	49.37	52.73	✓
1987	49.02	54.04	51.55	52.15	55.48	✓
1986	39.90	37.80	43.30	50.40	53.35	✓
1985	73.50	80.70	81.10	79.85	80.00	✓

Settlement on day nearest day listed.

Prices in ¢/gal.

Tactical variations

The best variation seems to be to buy on March 1 and sell 50% on May 1 and 50% on May 15. But any system can be used that buys in early March and sells later. Selling can begin any time after April 1. The profit and loss figures for the March 1 to May 15 period, taking one-half profits on May 1, are shown in Table 11–6.

Table 11–6 June unleaded regular gasoline—tactical trade variations

Year	3/1	5/1	5/15	Average	P/L	Result
2003	103.39	79.02	86.73	82.88	-20.51	✗
2002	71.26	80.48	78.82	79.65	+8.39	✓
2001	85.19	105.84	100.36	103.10	+17.91	✓
2000	88.10	83.27	96.79	90.03	+1.93	✓
1999	40.03	55.19	52.00	53.60	+13.57	✓
1998	52.15	54.29	50.75	52.52	+0.37	✓
1997	61.63	62.63	64.38	63.51	+1.88	✓
1996	57.66	67.42	67.20	67.31	+9.65	✓
1995	55.89	64.15	64.90	64.52	+8.63	✓
1994	46.83	50.47	51.17	50.82	+3.99	✓
1993	60.36	61.81	59.12	60.46	+0.10	✓
1992	60.15	64.69	63.68	64.18	+4.13	✓
1991	61.67	70.83	69.08	69.95	+8.28	✓
1990	62.52	61.03	63.51	62.27	-0.25	✗
1989	53.59	72.61	66.09	69.35	+15.76	✓
1988	45.27	49.37	52.73	51.05	+5.78	✓
1987	49.02	52.15	55.48	53.81	+4.79	✓
1986	39.90	50.40	53.35	51.87	+11.97	✓
1985	73.50	79.85	80.00	79.92	+6.42	✓

Settlement on day nearest day listed.
Prices in ¢/gal.

Using that system rigidly, there would have been profits in 17 out of 18 years for a net gain of $1.2330/gal, or $51,786.00 per contract, before commissions and fees. That does not include 2003, which brings the trade down dramatically. It was obvious at the time, though, that it was not going to be a seasonal year.

The average (without 2003) is a net gain of 7.25¢/gal per year, during the years it worked. During the last 6 years, it worked rather sensationally twice. The trade yielded double-digit profits in 4 of the 16 years that it worked. The only other year it did not work would have resulted in a loss of 0.25¢/gal. This plan had worked 12 years in a row before 2003. But, there was far too much risk that year, and its use was not recommended to anyone.

CRUDE OIL

Crude oil prices seem to have the same kind of seasonal tendency as the refined products have. During the last 21 years, it would have been profitable in 14 years to buy October crude in July. In 3 years, it would have been unprofitable. In 4 other years, one could have made a profit if one had been nimble. Those 4 years are considered *washes*, because it would have been easy to miss out on profits unless one accurately judged the highs. Still, there was a tendency for prices to advance from early July even in those 4 years.

That makes the buy side of the trade reasonably easy to make. Profits should be taken as they appear, anywhere from early August until the contract goes off the board. This seasonal plan gives guidelines concerning when to buy, but when to sell must be played by ear (see Table 11–7).

Table 11–7 NYMEX October crude oil futures

Year	7-01	7-08	7-15	8-01	8-15	9-01	9-15
2003	29.45	29.41	30.50	31.95	30.99	29.41	28.14
2002	26.48	25.87	26.70	26.06	28.38	27.79	29.67
2001	25.81	26.87	25.91	26.13	26.74	27.20	28.81
2000	30.20	30.28	31.40	27.58	30.96	33.38	35.92
1999	19.28	19.76	20.44	20.53	21.52	21.99	24.13
1998	15.18	13.85	14.87	14.50	13.35	13.73	14.57
1997	20.09	19.88	19.83	20.31	20.26	19.65	19.61
1996	19.86	19.79	21.06	20.48	21.40	22.25	23.19
1995	16.95	16.86	16.98	17.45	17.23	18.04	18.92
1994	18.65	18.57	19.04	20.11	18.19	17.47	16.70
1993	18.97	18.30	18.12	18.25	18.30	17.97	16.86
1992	21.70	21.35	21.54	21.77	21.22	21.64	22.18
1991	20.61	20.96	21.21	21.19	21.43	22.26	21.83
1990	17.94	17.90	20.33	22.10	26.36	27.32	31.76
1989	18.95	18.97	19.09	17.92	18.29	18.85	19.96
1988	15.27	15.79	15.08	16.20	15.81	15.08	14.90
1987	20.08	20.32	20.99	21.03	20.29	19.63	19.70
1986	12.28	11.42	11.38	11.38	15.91	16.46	14.34
1985	25.69	25.84	25.86	26.61	27.57	28.08	27.92
1984	29.90	29.80	29.39	28.38	29.36	29.23	29.28
1983	31.20	31.20	31.64	32.12	32.05	31.60	31.48

Settlement on day nearest day listed.
Prices in $/bbl.

PSYCHOLOGICAL MARKETS

Every now and again, there is a psychological market, ordinary at its beginnings and completely out of control at its end. These markets follow no rules or logic until long after they have ruined many people. They are also called panics, manias, crashes, or bubbles and are seen in a large variety of commodities, including oil and natural gas. Psychological markets can confound the wise and the foolish alike.

Psychological markets create a sense of hysteria. When a bubble is in full tempest, it prohibits a person's ability to remain detached and see the market objectively. It becomes a market Macarena—ridiculously popular one day and then cast aside without warning the next.

One obvious example occurred during the tulip mania of the 17th century. The tulip was first noted in Western Europe in the mid-16th century. In the early 17th century, the tulip gained in popularity. By 1634, it was considered an indispensable social prop that no man of wealth or taste could be without. Many had full collections of tulips at the time. The desire to collect tulip bulbs became such a rage in Holland in 1635 that price of a tulip bulb reached overwhelming amounts (see Table 11–8).

Table 11–8 Price of one Viceroy tulip bulb

Commodity	Value	Note
2 lasts of wheat	448 florins	A last was 80 bushels
4 lasts of rye	558	
4 fat oxen	480	A florin was two shillings
8 fat pigs	240	
12 fat sheep	120	
2 hogsheads of wine	70	A hogshead was 63–140 gallons
4 tuns of beer	32	5376 12-oz. glasses!
2 tuns of butter	192	1 tun equals 252 gal
1000 lbs of cheese	120	
1 full bed	100	Beds were left in people's wills
1 suit of clothes	80	
1 silver drinking cup	60	
Total	2500 florins	All for a single tulip bulb![†]

[†] Mackay, Charles. *Extraordinary Popular Delusions and the Madness of Crowds.*
(New York: Farrar, Straus & Giroux) 1932.

To try to put this in perspective, four tuns of beer is the equivalent of 5376 12-oz glasses, or close to 2½ six-packs per day for one year, about 224 cases. People pretty much only drank alcoholic beverages, watered down beer or wine, with meals. This started at a very young age, but that was hardly an excuse for the tulip craze.

The same kind of thing is seen periodically with pyramid games that occur every 10–15 years and gain legions of zealous acolytes in very brief periods of time. It requires restraint not to jump in with everyone else, at least the first time. A person who shows restraint begins to feel discouraged when later converts are still making money—*free* money. It is one of the most irresistible lures in life, but then it ends. It always ends, because there is never an inexhaustible supply of willing players or buyers, not for pyramid games, tulips, stocks, or pet rocks. The same is true for heating oil or natural gas in the heart of winter. Some years, the battle for supply is intense, but then the weather warms up.

Every other year at Christmas time, some unheralded doll or action figure or toy becomes *the* must-have gift for every child in America. The parents do not know it is coming, the manufacturer does not have a clue, and heaven only knows how every child develops the same urgent and spontaneous desire. This trend is picked up by news groups near its ending, but kids do not watch the news much. Sometimes the item is marketed heavily, and sometimes it is not. The knee-jerk reaction is to blame it on modern mass media, but if grown-ups had to have tulips nearly 400 years ago, it seems a bit shallow to eagerly embrace that theory.

Mood rings, black lights, Cabbage Patch Kids, clogs, Teen-age Mutant Ninja Turtles, Power Rangers, "Tickle-Me-Elmo" dolls, and little stuffed animals called Beanie Babies were all examples of these must-have *items.*

The first appearance of an item that will lead to a craze cannot be pinned down. They seem spring up spontaneously, appear everywhere at once, and then vanish just as suddenly.

Only 10 years before the Civil War, men were almost universally clean-shaven. It is not easy to pick out Robert E. Lee or Stonewall Jackson from pictures taken of them only a few years before the war began. It was not just the rigors of campaigning, because the same bushy faces are not seen in photos of the Mexican War. But during the Civil War, facial hair for men was wildly popular. J. E. B. Stuart and James Longstreet had beards of decidedly biblical proportions. It would be interesting to see if 10 experts could pick out computer-generated portraits of these men without their whiskers.

General Ambrose Burnside had such outlandish sideburns that the name for them came from his last name, turned inside out.

The same thing happens in markets. In the first Gulf War, everyone recited the mantra, "As soon as the shooting starts, all bets are off. The market could go to $50.00 a barrel!" When the shooting started, prices did rally $7.00/bbl early in the evening in London. However, by the opening in New York some 14 hours later, prices were down $7.00/bbl, for a net loss overnight of almost the next day's price for a barrel of oil. What everyone wanted at $32.00/bbl before the shooting, no one could give away at $20.00/bbl less than 50 hours later. The fear of war was much worse than the actual event.

Wars are frequently the cause of market runups, but they can just as often usher in bearish moves. More often than not, bull and bear markets seem to occur because they are meant to. For every purpose under heaven, there is a season. This is what W. D. Gann believed, and he may have been the most successful trader of all time. Shortly before he died, he drew a chart of what the stock market should do during the following 12 months. With a few minor deviations, he called all the major highs and lows and caught every intermediate and major swing that occurred in the year after his death. Legions of avid students are still trying to determine how exactly he accomplished this.

Trading Adages, Good and Bad

Wall Street is filled with its own history, legends, and lore, and it is inhabited by more than just bulls and bears. Clever snakes lie in the tall grass waiting for the right opportunities, and there are gluttonous pigs that, one is assured, eventually will be slaughtered. Sly old crocodiles pour out hot tears to lure one into dropping one's guard, and ravenous sharks gobble up companies and industries whole. There are animals that hunt in packs or prides, and there are the great herds they can frighten into stampeding. Lone eagles have keen eyes, hunting while they ride the wind. Ostriches hide their heads in the sand, and deer freeze in the headlights.

The markets are filled with allusions to animals. A stock or commodity that refuses to move higher with others in its complex is called a *dog*, while one that advances beyond its peers is considered a *high flyer*, or a *bellwether*. A *wether* is a male goat, and goatherds used to take the dominant male in a herd of goats and tie a bell around his neck. Since the other goats were used to following him anyway, it was a good way to keep track of the goats. Sheep would often follow the goats as well, and market sheep often follow the bellwether stocks or commodities.

Like all of the animals mentioned, there are living examples in the markets. Different strategies work well for different types of traders. The same is true for market adages, sayings that have been passed down through the generations as trading lore. Some of them are nonsense, designed to lure new traders into mistakes. Others are pearls of wisdom, and still others are throwaway statements made by successful folks who were good with what is now called a sound bite. Some of these adages will be discussed here.

"You never go broke taking a profit," or, "Let your profits ride and cut your losses short."

Both of these sayings are true. In terms of the market and trading, however, the better adage is Honest Abe's 1864 reelection slogan, the second of the two quotes above. It is wise trading advice to "let your profits ride and cut your losses short."

Commodity prices experience wild gyrations at inconvenient times. Until those times come, though, there can be long and boring periods when nothing much happens. Prices may be in a range, or they may *fade* their openings (move higher from a lower opening or lower from a higher opening). It becomes easy for a trader to find a certain predictability in the rhythmic movements of a market in this condition.

However, it never lasts, and the change always comes at the worst time. There was once a sugar trader who was given permission to trade company funds, and he started off brilliantly. For a couple of days, then for several weeks, he made steady profits every day. When asked how he did it, he smiled and said, "Easy, I just never take a loss." It sounded too glib to be more than bravado, but when pressed, it turned out to be the undiluted truth.

The market was trading quietly in a range, and he would sell near the upper end of the range and buy near the lower end. If it went a few cents against him, he would hold on until it swung back the other way. He waited for each position to show a profit before closing it out. He had been deadly serious; he just never took a loss, at least not up until a point that came later.

Of course, this trading technique did not continue to be profitable indefinitely. One day, the trader was short 20 lots, and the market ran up 50 points, after having had a daily range of 20 points for weeks. He decided to stay with the plan and held on. By the time he got out, the market was 200 points higher. He eventually lost three times more than what he had made in the many weeks in which he had been right.

They say that generals are doomed to fight each new war with the tactics of the previous one. And in most cases that seems to be true. It is especially true in trading, where the rules change very quickly. When prices move from a trading pattern (in a range) to a trending pattern (moving in a direction), traders need to adapt faster than battlefield generals. The mistake the sugar trader made was that he did not adapt quickly enough to changing market dynamics.

The game of rock, paper, and scissors can be used as an example. Normally the paper covers the rock, the scissors cut the paper, and the rock crushes the scissors. The situation would be different, however, if there was a referee who could change what object beats the other object without letting the contestants know the new rules. Nonetheless, with two possibilities, a person still could figure out what the rules were pretty quickly. This would be true even if the referee continued changing the rules from time to time, as long as he did not change the pattern too often. A mistake would be informative, and a player might even take notes.

People are allowed to do as much in the market, and it is important that they change tactics as the trend changes. When in doubt, it is better to preserve resources for another day. The goal is not to fight just one battle but to fight a campaign or a

Adapt to Changing Tactics

If a frontal assault worked well for Napoleon but not for Robert E. Lee, why did Douglas Haig try it? The rules had changed (infantry weapons fired faster and were accurate to longer ranges, there were machine guns, the tactical reduction of shock troops, and the effects of massed artillery). Most of these changes had begun in the American Civil War. Cavalry became a scouting and screening arm, and frontal attacks against an entrenched enemy led to the carnage of Marye's Heights, Pickett's Charge, and the Bloody Angle at Spotsylvania. But Douglas Haig refused to learn, even in the face of devastating losses in World War I. Millions perished as Great War generals failed to learn from events that had occurred in the American Civil War and in the Crimean War. The charge at Balaclava, the "Charge of the Light Brigade," was romantic poetry, but it also showed how miserably a frontal cavalry charge did against modern firepower. It took World War I generals three bloody years to learn from earlier mistakes and from their own staggering losses. They were very slow learners.

war. If losses become higher than usual, it should be taken as a sign that something has changed. One should pull back, husband one's resources, and wait to trade another day.

Common sense is vital in trading. Higher losses than normal should also be taken as a warning that a person should stop and reevaluate his trading approach. The rules, or the trend, may have changed. High losses are *always* a bad sign. Trading capital should be viewed as no less important than a person's health. If someone throws up blood every time he enters a pie-eating contest, he would stop entering the contests and likely would call his doctor. In trading, people will be losing cash daily after trading too many contracts in the S&P market, but they still will not stop

It might be helpful to consult an accountant, an analyst, or a financial professional. However, it may not be the wisest move to consult a broker, especially if that broker recommended the trade in the first place. Brokers sometimes get emotionally attached to their customers, their customers' positions, and the commissions that their customers generate. This is said without any insinuation intended, because a broker *does not* want to see a client lose money.

The broker who recommended the position in the first place will have pride, as well as money, on the line. Furthermore, the broker may have recommended the position to other customers, as well. The broker may have many other customers who also are holding losing positions.

Consequently, a person in this situation discovers that when he most needs the advice of his broker, the broker will be particularly distressed and distracted. In reality, the situation is many times worse for the broker. He feels the agony of his clients' losses, the pressure of asking for margin calls, and the insecurity of losing a livelihood. Invariably, he will advise holding the position, in a gloomy voice that abandons all hope at a significant top or bottom. Someone who likes and respects his broker will find that a word of encouragement offered at those times will earn more than might be imagined.

A person with a losing position should ask his broker if he would recommend *adding* to this position where it is. If he does not recommend adding to the position, then it is being held only in the hope of a favorable reaction. In this case a person should go back to the start and consider carefully for himself if it seems a good position to enter now. If it is not, or if he is in doubt, he should get out.

Situations such as this reveal why it is preferable to use brokers who do

If one is going to be a trader or a broker, stress will be the biggest worry. Clinical studies show cats and dogs help to relieve this stress. Experience has shown that nothing works better. The last thing one needs is to talk to someone who has many of the same fears or hopes for prices. Pets offer better emotional support than brokers or traders, and they are not going to rattle a person with new and contradictory information or old and discounted theories. This is not meant to be disrespectful, but often traders want someone to provide comfort or hold their hands. This only further distracts everyone involved from finding solutions.

not trade their own accounts. That way, if their recommendations go badly, they will not be distracted by their own losses when advice is needed the most. In any event, it is wise to be aware of the stress level that is likely to be felt by one's broker on any given trade, should it turn out badly.

Successful traders seem to follow derivations of the rule to limit losses and let profits run. With the volatility in today's markets, *probing* the markets can be a profitable way to trade. A predetermined stop point is set where a person trading will admit he is wrong. But winning trades are held, following them with a *trailing stop*.

One rule of thumb is to enter only those trades where the objective is at least three times the risk. A person who buys heating oil at 50.00¢/gal, and has a stop at 49.00¢/gal, should have a strong feeling that it can go to 53.00¢/gal. In this manner, he can be wrong three out of four times and still break even. A person who is right one-half of the time will make money, at least in theory, but few carry it through in practice. Although it is a good system, it requires patience. If followed diligently, there will be occasional windfalls that can be huge if the stops are adjusted to reflect higher volatility.

Strictly speaking, no one ever went broke from taking a profit. But, as a trading rule, it can lead to ruin. It is better to take a disciplined approach and to let profits accumulate while cutting losses short.

"Buy low and sell high."

This is one of Wall Street's favorite expressions, and it brings to mind the freshman asking the varsity coach if the school football team will go undefeated this year. The coach does not miss a beat, answering, "Yes, if we win them all." It sounds great: buy when it is low and sell when it is high. However, it is not that easy. How does one know when it is low or when it is high?

More often than not, it is better to follow the trend. And that would change this expression from "Buy low and sell high" to "Buy new highs and sell higher, or sell new lows and buy lower." The markets are filled with people who have no idea of values but too many ideas about price. They sometimes try to tie them together, but the two make a bad bundle. In commodities, everything has a price that is easy to see and a value that may be next to impossible to discover.

"You've got to be in it to win it."

This is a very catchy theme for a $1 or $2 bet on the state lottery, but of course, those risks are very different from the risks associated with futures. A person who likes state lotteries may love options. And one can play the state, too, if one likes (by selling or writing options).

This adage does not take into account that there are three positions in the futures markets: *long, short,* and *flat.* It fails to recognize that being on the sidelines is a legitimate position in the markets. If state lotteries raised people's property taxes if they did not get any of the numbers right, people would see the sidelines as a legitimate position there, as well. "If you're not in it, you can't win it," is a true statement. But it is also true of the heavyweight boxing title.

"The trend is your friend."

A disarmingly simple saying that is worth its weight in gold, or any other commodity. One can spend years examining the markets, trying to analyze the fundamentals to the smallest detail, and one would still have a hard time coming up with a more useful axiom. The frighteningly simple truth is that trends continue more often than they change. A person who sticks with the trend will have a much better trading career.

"You can always get back in again."

This is true only in the sense that divorced couples can always get back together again. At the heart of this adage is a boundless optimism that is capable of ruining weeks of diligent work. If a traveler was flying to San Francisco from New York, and the plane stopped in Kansas City, would he get off in hopes that he could link up again with a flight from St Louis?

That would be unlikely. A person who is trying to get somewhere has a plan and sticks to it as closely as possible.

Invariably, there are traders who see a big move coming but exit their positions prematurely. A trader who is riding a trend does not get out just because he has been in it for a while. Chances are he will miss an even bigger move in his direction and then need to fight to get back in. It is true that one can always get back in, but this should never be used as a reason for getting out. One should have a real reason.

"Bulls make money, bears make money, and pigs get slaughtered."

This vague statement sounds good, but it would be difficult to pinpoint its meaning. This might be some oracle's way of saying not to be greedy. This does not have many useful applications.

"Anyone can make a million dollars trading. It's holding onto it that is the trick."

This saying is based on our simple observation that those who are bold enough to make a lot of money are generally not conservative enough to keep it. Those who are conservative enough to keep it are generally not bold enough to make it. If these two types are put together, they will be bold enough to lose everything they have and conservative enough not to get it back.

"The market always moves farther and faster than one would dare to predict."

This saying of ours is based on years of observation. Those few who make large profits have found a way to factor themselves almost completely out of the equation. They do not get caught up in the minute-by-minute gyrations, and they do not get caught up in the psychological maelstrom of the market. If one is smart enough to buy at the bottom, it will require every ability one can muster to keep from getting out too early. The definition of a bull or bear market is its ability to stun people and its capacity to surprise with its stubborn virulence.

"People make money in their own businesses, and they lose it in other's."

So many times people succeed in the business they know, only to lose in a business they are not familiar with. We once asked a client how he had accumulated 50 gasoline stations. The first made a profit before he bought the second, we conjectured, and the first 15 made money before he bought the next 10, we guessed.

And yet he approached trading differently. He cashed in profitable trades and held on to losing trades. That is not how his family got their business started, and it is not generally a good way to start speculating. The lessons one learns in business can be adapted to trading, and the most prominent of these is the admonition not to invest more in losing propositions but to add only to strategies that have succeeded.

Corners, panics, and squeezes

A squeeze can turn into a panic, which is what happens more often than not in attempted or rumored corners and squeezes. Bunker Hunt almost cornered the silver market in 1979, but when all else failed to stop its rise, the shorts forced a temporary rule change through the COMEX board. Bunker Hunt was left holding more silver than he could finance.

Squeezes almost never succeed. They are like a game of chicken with cars and a cliff up ahead. In order to corner a market, someone has to have a position that is too large to get out of gracefully. Anyone who cannot get out in time is going to go over the cliff. To top it off, they usually lose their dignity, too, trying to claw their way out after it is too late.

These are descended from the pools that many large speculators ran 80–100 years ago, before anything was illegal on the exchanges. Large traders would accumulate a stock with other members in a pool of traders. Once they had accumulated a full line of positions, they would start *bidding her up*. Typically, this would be in stocks. They would have enough money in the pool to lay down one-half interest on a block of say 25,000 shares (when that was a large number). They would have their broker wade in and announce a bid for 25,000.

Since there was a rule forbidding partial fills, these orders would scare out the shorts and frighten other potential buyers into believing that they needed to buy right away. Any holdouts were, in turn, persuaded by the buying of the first crowd. And this process would be reported without the longs needing to buy many more shares than they had already accumulated at the lowest prices.

The price kept advancing until there were no more shorts left to cover or would-be buyers waiting to buy. At that point, the pool would generate a sell-off to attract some fresh bears and to scare out some late-arriving bulls. They would repeat the process until there were no more bears willing to sell short or stopped-out buyers looking to get back in. The pool was then long at the top.

Of course, getting out was a bit more difficult. Just before everyone realized what was occurring, the organizers of the pool were supposed to be selling out their longs behind the backs of the others in the pool. This did not always work, for the smaller the amount that a person has, the more jealously he guards it. And it was in this process of doubt and greed that the squeezes or corners often came unraveled. Individual greed will generally overcome pooled or shared greed.

Are bad fills really bad?

A bad fill occurs when a person pays above the last market price to buy or when a person sells at a lower level than expected. Bad fills, in this case, occur on market orders or stop orders. I always liked getting bad fills when I was initiating positions. The worse the fill, the happier I would be. Some traders get upset over bad fills, but I never had a bad fill that was not a winning trade—the worse the fill, the bigger the profits.

How could anyone like a bad fill? Bad fills signal that a person is swimming with the current and has hit the rapids. If one puts in a limit order to buy and gets filled well below it, those extra points will be scant solace when one makes one's margin call the next morning. If it is that easy to buy, it means that the selling is ferocious. In traders' heaven (if there is such a place), maybe one can get great fills and still catch the big move. But in reality, a great fill means that one is driving on the wrong side of the road.

Of course, getting out of a position is a different story. Even so, a person who gets a miserable fill getting out of a position may take it as a good sign. The situation is likely to get worse, and it was probably wise to get out.

The only thing better than a bad fill (when initiating a trade) is a partial fill. Partial fills are as close as a person can get to a guaranteed winning trade, especially the longer they remain only partially filled. It is not a difficult concept to grasp. By its very definition, a partial fill is a winning trade, at least at the time it is reported, and thereafter as long as nothing new is reported. If one wanted to buy 25 lots at 55.00¢ and got 10 of them, it means that the sellers were not good for another 15 at that price. If it stays only partially filled for 10 minutes and then starts to move up, the remaining 15 contracts should be taken to the market. The price is not as important as where it is going next.

This does not mean that one should hit locals' bids or lift their offers on out-of-the-money options on a contract 10 months away. Really, one should not even be trading if it can be avoided. One should at least go to the at-the-money option two or three months out or beyond. And that is where one wants to use limit orders.

Splitting the bid and offer is our preferred approach. On any active contract, we like going to the market. If one feels prices are going higher, one should buy it. It is not wise to muck around trying to pick some local's pocket over the phone for the sheer satisfaction of it. Just as often as not, one will miss the boat by getting cute with actively traded contracts.

One should not try to impress one's friends by trading over lunch or during a round of golf.

The market has it in for these traders, and it derives a special joy from humbling the arrogant. Actually, the market seems to humble everyone, but it appears to be the proud who really get plucked. It would not be smart to take a hot new date to the restaurant where one's spouse works, as a wise precaution against certain calamity. The market is just as jealous. Many jobs have been lost by people who took the market for granted.

When one is trading, one should trade. When one is doing anything else, one should not trade. It is like drinking and driving—they do not mix. If one is a chauffeur, it is especially inexcusable. It obviously would not be wise to drink and trade. Neither is it wise to trade while driving or playing a round of golf. Trading never helped anyone's golf game, either.

It may seem impressive to brag about making money with a cell phone call on the tee-off on the fifth hole at some swanky golf course. But as far as trading goes, it is courting disaster. The market should be treated with respect.

Is there such a thing as real value?

Every now and then someone may consider the curious notion that a commodity has an underlying intrinsic or real value. A person who feels that way after buying something that has already dropped in price should get out of it before the feeling further contaminates his thinking.

Everything in the commodities world has a price, but nothing has a real value. What is the real value of a tulip bulb? It is safe to say it is not 2500 florins. And yet people were willing to pay that price in 1635. In June 1990, crude oil was priced at $15.06/bbl; by October, it was at $41.15/bbl. Somebody bought and sold at least 1000 bbl at both prices that year. What was it really worth? It is hard to tell, and asking that kind of question is dangerous. Finding, getting, or having an answer is even worse.

Sure, comparative analysis is useful, because people get used to certain price levels, and they translate those prices into a concept of value. But it will get in one's way here, especially if it is in any way used as a reason for keeping a losing trade. One should not rationalize in this manner.

"Your first loss is your best loss."

This is a good piece of advice, and it can be helpful for those trading long-term or short-term. An increment should be established, either in price or in time. If prices fall below this mental sell-out point (or mental buy-back point if short), or if at the end of the allotted time period the trade is losing money, the position should be exited. There are plenty of intermediate trades with which a person can comfortably use the end of the trading day or the end of a three-day period or a week as the deciding point. The person who finds the market more than 20 points against him should get out and take a fresh look the next day.

Employing this system or thought process can prevent those really debilitating losses that can ruin any trading program. It will take time to get comfortable with a money or time amount that fits. But, once a rhythm is discovered, just its practice can help.

During my years as a broker I handled many accounts. And the two biggest reasons that people lost money were overtrading and a refusal to take a loss before it became too big to get out of gracefully. As soon as it passes that point, it is like a lame horse. The loss gets bigger, but it has already passed the point of being "too big a loss to take," so the trader continues to sit in horror as the situation worsens. Multiply this over dozens or hundreds of traders, and the result is a big pile of limit orders that need to be lowered several times without getting filled. At the end, the traders get out at the point where prices turn around. It is just one last insult heaped on a pile of injury.

This will help the reader to understand what is meant by the next statement.

Markets always move in the direction that will hurt the most people the most.

The markets seem to instinctively move in a direction that will cause the greatest pain to the greatest number of people. That is perhaps the main reason to follow the "Commitments of Traders" reports so religiously. They can reveal where there can be stampedes or bottlenecks if prices get up a head of steam.

Inevitably, at the major highs and lows, there are lots of smaller traders holding a smaller number of contracts, and they are ranged against a few traders holding very large positions on the other side. And as night follows day, eventually this disproportion catches up with the smaller traders.

A person will develop or collect his own useful axioms as time goes by. It is vital to learn from mistakes as well as successes. One way to do this is to keep a market diary and write down what worked and what did not, periodically revisiting what has been written.

Glossary

Access. An electronic trading platform used for trading oil and natural gas prices after the regular NYMEX session has finished.

accumulation. The consolidation period at a bottom. At this stage, strong hands are accumulating long positions in anticipation of a move higher. The number of individuals holding long positions drops, while the average size of the long positions increases. Long positions become concentrated in fewer but stronger hands.

ADP. See **alternate delivery point**.

alternate delivery point (ADP). A cash-based way for two parties to exchange physical barrels at a mutually convenient delivery point not specified in the NYMEX delivery specifications. Both parties have the security of trading on the NYMEX. Cash transactions are only as secure as the two counterparties.

any month barrel. A barrel to be delivered at any time during a month. NYMEX futures contracts lock in prices and call for delivery of any month barrels, or barrels for delivery at any time during a specified month. Delivery is made at the seller's discretion.

artificial risk. Gambling. The creation of risk for bettors.

ask. The best (or lowest) selling price in the market at any given moment. Also called an *offer*.

assuming risk. What speculators do. Taking on risk in order to potentially make a profit. Speculators assume risks that commercial enterprises want to lay off.

at-the-money. The closest priced option to the existing price, or one that is at breakeven. This also applies to capped price programs, in which a cap acts like a call option.

basis. The difference in price between one's local rack prices and the NYMEX price.

basis risk. The risk that exists between the price of futures and the actual cash price one must pay. See also **time basis risk** and **location basis risk**.

bbl. The common notation for a barrel of oil. Each barrel is 42 gallons.

bear. Someone who is short or believes that prices will move lower. This usage of the word *bear* comes from an old adage, "He went out and sold the skin before he shot the bear." Since short sellers sell things they do not yet have, they are called bears.

benchmark. A type of crude oil against which other crudes are priced. There are currently three: West Texas Intermediate (WTI), a light, sweet benchmark crude in the United States; Brent Blend in Europe; and Dubai in Asia. Other types of crude oil are generally priced against these benchmarks.

bid. The best (or highest) buying price in the market at any given moment.

bpd. The common notation for barrels per day.

broker. Someone who clears or executes a trade. A broker represents clients and is licensed to make bids and offers on the phone to the floor or upon the floor of an exchange.

budget or **budget customers.** Heating oil customers who pay for their winter needs over the course of 10 months or a year to spread out the payments. Typically, the barrels for these customers will be hedged so that the monthly price they pay can be set ahead of the actual use.

bull. Someone who is long or believes that prices will go higher.

call. Option to buy a futures contract at a specified time and price.

capped price programs or **caps.** Programs offered by suppliers, often refiners, that allow users to pay a premium up front to lock in the highest price they will pay, without denying themselves the potential benefit of paying less if prices decline.

cash prices. The prices one pays for physical product, typically in one's own region or area. Cash prices are those one pays (or gets) to take (or to make) actual, physical delivery of a commodity.

cetane. A measurement used in distillates similar to octane in gasoline.

Colonial Pipeline. The major pipeline carrying refined products from the U.S. Gulf to points along the East Coast. The underlying cash instrument used to determine NYMEX specifications.

commercial traders. People who buy and sell cash and/or futures because they make or use the underlying commodity. Commercial traders represent businesses involved in the purchase or sale of the real commodity.

commission house. A brokerage house.

contract. A financial instrument that calls for a specific commodity to be delivered or taken delivery of at a specific point in the future. Contracts are designated by months and are interchangeable. The sale of one contract offsets the purchase of another contract of the same commodity in the same month.

differentials. The price difference between one commodity and another related commodity, or one location and another location's price, or between one delivery time and another delivery time's price.

differential risk. The risk that one related commodity price will move more dramatically than another, that one time period's price will move dramatically against another time period's price, or that one location's price will move dramatically against another location's price.

directional risk. See **price risk**.

discount. When one related price is beneath another price, or when a differential is beneath the figure to which it is being compared.

distribution. This is where big traders sell their large holdings to smaller traders in smaller parcels. More individuals become long, but their average position becomes smaller. Long positions are moved to weaker hands.

downside potential. The potential for prices to move lower.

EFP. Exchange for physicals.

end-users. Commercial entities or individuals who actually use a specific commodity. The people who consume or use the product.

expiration. Both futures contracts and options on futures contracts have specific days upon which they expire. Once a futures contract (designated by a specific month as its name, like January, February, etc.) expires, no new trading is possible, and those still holding contracts must dispose of them through EFP, ADP, or physical delivery.

fade. When one *fades* the opening, one is going the opposite way. In other words, one would fade a lower opening by buying it or fade a higher opening by selling it. It is most commonly used with advice about opening levels.

Fibonacci. A Renaissance mathematician who rediscovered and reapplied a naturally existing ratio that can be seen in sunflower seed placement and in the curves of a snail's shell. This was used by the ancient Greeks in architecture. It is used in technical analysis and is based upon the ratios of 0.618 and 1.618, which are each others' inverses. The Fibonacci sequence of numbers starts with 1, adds it to itself to get 2, then adds 1 to 2 to get 3, 2 to 3 to get 5, 3 to 5 to get 8, 5 to 8 to get 13, 8 to 13 to get 21, etc. Each set of numbers is in the ratio of 0.618 or 1.618.

fill. The price returned from the floor to the customer. It is a trade that has found a buyer and a seller willing to agree on the same price. An unfilled order is one where a would-be buyer or seller has not found a match yet.

fill or kill (FOK) orders. Perhaps the price is trading at 52.95¢, and one wants to offer a contract for sale at 53.00¢ but does not want to be out there all day with one's broker always offering to sell the contract. In this case, one might give him an FOK (fill or kill) order to sell at 53.00¢. One's broker then goes out and offers the contract at 53.00¢ three times. If no one fills his offer, the order is automatically canceled.

fixed-price programs. These programs allow end-users to lock in specific future buying prices for refined products. These programs represent physical barrels at a fixed, specific price at a specified location during a specific month. Because they represent wet barrels, they can significantly reduce basis risk between the NYMEX and the terminal one uses to get physical product. The greatest drawback is that one is obligated to pay the fixed price regardless of where prices are when one picks up product. If prices drop, one must still pay one's fixed price. The greatest advantage is that one will never pay more, regardless of how high prices are, either on the screen or at one's terminal. One is protected against price increases.

fundamentals or **fundamental analysis.** The practice of trying to determine future price trends by looking at the supply and demand factors in the market. Using fundamental analysis, one is trying to gauge and discount the reasons for a price move. This is different than technical analysis.

funds. See **trading funds.**

fungible or **fungibility.** One is identical and interchangeable with another. Futures and options contracts in the same month (with the same strike price in options) can have purchases and sales offset each other. Every futures contract (with the same month) is fungible with every other futures contract of the same month. They represent exactly the same underlying commodity. One crude oil contract of 1000 bbl for delivery of light, sweet crude at Cushing, Oklahoma is the same as any other.

futures. An obligation to make or take delivery of a specified amount of a specified commodity at a specified location at a specified time in the future. Futures contracts are designated by the month in which delivery is expected to take place. Futures purchases may be offset by futures sales in the same month of the same underlying commodity.

good 'til cancelled (GTC) orders. This allows one to leave in an order for more than just the one-day session. Unless specified GTC, every order expires at the end of the trading day. One should be careful with these and check with one's broker periodically to see what outstanding orders one may have if one uses these orders. There are plenty of examples of traders who have put in GTC orders only to find themselves getting filled in them when they least expected to be. Frequently, traders will place these *wish* orders and forget they have them out there. If and when they do get filled, it will be in the middle of some unfortunate move like the ones that occurred in the winter of 1986, in December 1989, or during the Gulf War. One should use these orders sparingly and make sure one writes them down! Brokers generally prefer to avoid these orders. Some firms do not accept them because of potential problems with them.

GTC orders. See **good 'til cancelled orders.**

handle. The first two digits of a price. In trading, buyers and sellers use the last two digits to indicate buying or selling. Traders will say, "25 bid at 50," dropping off the first two digits, or handle, of the price.

hedging. The laying off of price risk(s) associated with doing business. A crude oil producer drilling for oil can hedge the price of barrels not yet brought to the surface by hedge selling future production, thus locking in a future selling price. An end-user can lock in future buying prices by hedge buying future needs. Hedging is the opposite of speculating. Neither is gambling.

inherent risks. The risks that automatically come from doing business. An airline has an inherent fuel price risk. If the price of its fuel is not hedged, a sharp rise in prices could negatively affect the company's competitiveness or profitability.

in-the-money. An option that is already making money. A call below the market price, or a put above the market price.

intrinsic value. Applied to options, the amount already "made" by the put or call. A call at a strike price beneath the current market or a put at a strike price above the current market is said to have an intrinsic value of the difference between the strike price and the existing market price.

lifting. Taking physical delivery. A buyer *lifts* barrels at a location.

lifting price. The actual price one pays for physical delivery. This can reflect time or location differentials, terminal fees, barge or tanker costs, demurrage, throughput, or pipeline costs.

limit orders. Orders placed to buy at a lower price or to sell at a higher price than the current market price. If one's limit is traded beyond, one is owed a fill unless the trade is taken down that violated one's order.

liquidity. A market condition in which it is easy to get in or out of an instrument without having to pay much more or receive much less than the previous market price. A market with strong interest and heavy volume has good liquidity.

locals. Locals are traders who operate on their own hook on the floors of the exchanges. They are considered professional speculators.

location basis risk. A type of basis risk. Futures call for delivery in New York Harbor. Barrels may be more or less available (and expensive) in locations removed from that delivery point. Barrels in Boston or Albany may be trading higher (at a premium) or lower (at a discount) in relation to New York Harbor.

long. Someone who has bought the market is *long* and is said to be holding a long position.

long hedge. See **upside protection**.

margin call. When one has a position that has lost money, eventually one may reach a point where one needs to bring the good faith money one has deposited back to its original amount. Since futures are traded on margins of 2% to 5%, small moves can generate large profits or losses. When one has losses, one gets a *grace* of about 20% of the original margin. Once it falls below this maintenance level, one will be called to refresh one's margin deposit to its original level.

markdown phase. This is the actual move lower. Prices are *marked down*.

market if touched (MIT) orders. This is very much like a limit order, except once one's limit is touched, the order becomes a market order. Perhaps one wants to buy December heating oil at 50.00¢. However, one does not wish to trifle with 0.05¢– 0.10¢ (5 or 10 points) and risk missing getting in. One could tell one's broker to buy one December heating oil at 50.00 MIT. If December prints 50.00 during the day, the broker will buy one contract at the next best available price. Normally, in an active heating oil market, that will be within 5 or 10 points.

market on close (MOC) orders. One's broker buys or sells (per one's instructions) on the close of the market.

market on opening (MOO) orders. One's broker buys or sells (per one's instructions) during the market's opening.

markup phase. This is the actual move higher. Prices are *marked up*.

mmBtu. The common notation for a million British thermal units, which is the per-unit price designation for the NYMEX natural gas contract, as well as for cash market prices. Every natural gas price is per million Btu.

MOC orders. See **market on close orders.**

MOO orders. See **market on opening orders.**

NYMEX. The New York Mercantile Exchange. The exchange upon which energy futures and options are traded.

OCO orders. See **one cancels other orders.**

offer. The asking price, or the price at which a trader is willing to sell a specified commodity.

one cancels other (OCO) orders. Perhaps one has a long position, for example, from 52¢, and one would like to sell it if it gets up to 55¢ but does not want to still be in it losing money if it breaks 50¢. In this case one would use an OCO order to sell one contract at 55¢ (on a limit) OCO 49.95 STOP. That way, one does not end up selling the one long contract twice. Professional traders will use these orders in much closer proximity.

open outcry. The method of trading used in commodities rings or pits. Buyers and sellers shout out bids to buy and offers to sell and are obligated to sell to the highest bid or buy from the lowest offer.

operating profits, operating margin, or **income.** The amount of money made through normal business operations. The amount needed by a normal business to remain profitable. This is in contrast to trading profits or windfall profits.

options. An instrument representing the opportunity without the obligation to buy or sell a specified amount of a specified commodity at a specified price, time, and location in the future. A call is the opportunity to buy, while a put is the opportunity to sell. Options may be bought by paying a premium, or they may be sold or written by someone potentially willing to take the obligation that comes from offering the opportunity.

out-of-the-money. An option that is already losing money. A put below the market price, or a call above the existing market price.

out trades. Trades that either the buyer or the seller do not recognize as having occurred.

peaking units. Utilities and large industrial companies will use diesel-run generators to produce additional air-conditioning on extremely hot days. These are called peaking units.

pits. The physical spaces where trading in commodities is conducted. Pits are what they call them in Chicago.

pour point. The temperature at which oil or oil products remain fluid. Below the pour point, the particular product will no longer flow unless it is cut with something that can lower its pour point.

premium. 1) The amount one must pay for an option. An option buyer only risks losing the premium. 2) The amount above cash or futures that one must pay to obtain wet product at a specified time or location. 3) A grade of gasoline with octane content in excess of 91.

price risk or directional risk. The risk that prices will move against one. A would-be buyer, say an end-user, has a price risk that the needed product will increase in price before he has actually purchased it. A would-be seller, say a producer, has a price risk that the product will depreciate in price before he can actually sell it.

prompt. For immediate, right-away, delivery.

prompt premium. The additional amount one must pay to get immediate delivery.

put. Option to sell a futures contract at a specified time and price.

rack. The apparatus that fills tank-wagons at a terminal. Prices *under the rack* are those that will be paid by distributors or truckers to wholesalers.

real risks. The risks that come from operating a business involving the purchase or sale of commodities. These risks can involve any company using any commodity in any way.

resistance. A level on a chart where selling is expected to develop.

resting orders. See **limit orders**.

rings. The physical spaces where trading in commodities is conducted. Rings are what they call them in New York.

scaled-down buying or scaled-up selling. Buying at lower price increments as prices drop or selling at progressively higher prices as prices rise. Generally engaged in by the trade or commercial accounts. They will buy one contract at a certain price, another at a lower price increment, another at the same price increment below that, and so on.

settlement. The price at the end of the day against which open positions are calculated for purposes of adjusting the account worth and margin requirements.

short. Someone who has sold the market is *short* and is said to be holding a short position.

speculative profits. The amount of money made by a speculator when prices go in the foreseen direction. The reason or justification for speculators to assume risks being hedged or laid off by those engaged in businesses with inherent risks.

speculator. A trader who assumes existing risk in return for the opportunity to make a profit. Speculators assume risks that hedgers already have and wish to lay off.

spot outage. A situation that exists when one or more terminals in a region physically run out of material.

SPR. The Strategic Petroleum Reserve.

spread. A position that has a long contract *spread* against a short contract. If one buys December heating oil and sells March heating oil against it, one has placed a December–March heating oil spread on one's books. The variations are endless. One can buy futures and buy a put, or sell a call. One can buy heating oil and sell gasoline or crude oil or even natural gas. Theoretically, as positions, they are less volatile than outright positions, but traders can lull themselves into thinking in terms of lower risk when that is not always the case. Margins are lower on spreads, so the bang for one's buck can be magnified doubly.

stops or **stop orders.** Orders placed to limit potential losses in an open position. They can be used to initiate positions but are used that way less frequently. Once a stop price is touched, the stop order is triggered, and it becomes a market order. If one buys heating oil at 52.00¢ and does not want to take much more than a 2¢ loss, one could place a stop order to sell the contract at 49.99. If 49.99 or anything lower is printed, the order becomes a market order to sell. If the low for the day is 50.00¢ or anything above that price, the sell-stop will not be triggered. If 49.95¢ or 49.90¢ is traded, the sell-stop will be triggered. One's broker must then sell the contract at the next best bid. If the best bid after a 49.90¢ print is 49.80¢, then one could be filled there, selling at 49.80¢. Stops are a mechanical means that allow one to get in or out of a market more efficiently. Stop orders expire at the end of any given trading session and must be placed again the next day if one wishes to still have them in place. One could also place GTC orders, but one must be aware that they are there until one cancels them or the contract expires.

strike price. The specified price in an option.

support. A level on a chart where buying is expected to develop.

swaps. Instruments underwritten by banks, brokerage houses, or other financially backed institution. These act as a combination between forward markets and futures markets. They are as strong as the institution that backs them.

sweet. A crude oil designation meaning a low sulfur content.

technical analysis. The attempt to forecast future price movement by analyzing the trend, chart patterns, support and resistance, and volume and open interest. Technical analysts believe that there is no way that any one individual can fully discount everything known by everyone in the market. They believe that the aggregate buying and selling of this combined knowledge is reflected by what has happened in the market's prices.

terminal charge. The cost of lifting or storing product at a nearby terminal.

time basis risk. A form of basis risk. If one owns an any month barrel through futures, one may not be able to obtain that barrel physically at the beginning of the month. When one needs product, one needs it. Prompt, or immediate delivery barrels, may cost more to obtain.

time value. That part of an option's premium determined by the amount of time left on the instrument.

trade. A trade occurs when a buyer and seller agree upon a price.

traders. People who buy and sell any commodity.

trading funds. Money managed by professional traders. This is often money pooled from several different sources looking for a better-than-average return by assuming greater-than-average risk.

trading profits or losses. The amount one makes or loses strictly through trading operations. These differ from regular business profits or losses.

under the rack. See **rack**.

upside protection. A hedge designed to protect distributors or end-users against having to pay higher prices. Also called a *long hedge*.

violated. When a limit order is not filled but was traded beyond. If one placed a limit order to sell at 52.90, and it trades 52.95, one is owed a fill unless the 52.95 price is *broken*, or voided, and is awarded to one at 52.90.

wet barrels. Real, physical barrels of oil that one can actually burn. These contrast to paper barrels, which are financial instruments.

wet barrel programs. Supply arrangements that a distributor or end-user can make with his supplier to provide a specified number of wet barrels at a specified price, time, and location in the future. These help contain time and location basis risks.

wildcatting. Looking for crude oil on one's own hook. Financing an exploration and drilling operation on one's own. These operations can result in large profits or *dry holes*, where no oil is found and large sums of money may be lost.

windfall profits. Made famous in the late 1970s as an expression used to describe the additional profits made by oil companies with large amounts of oil in the ground as prices rose during the political turmoil of that decade. In modern usage, the term is used to describe profits that accrue without one doing any additional work. *Found money.* As used in this text, it means profits made by companies who do not hedge but get lucky because prices may happen to go their way. This is in contrast to *operating profits*, or the amount one might expect to make by running a well-organized and responsible business enterprise.

Index